SpringerBriefs in Computer Science

SpringerBriefs present concise summaries of cutting-edge research and practical applications across a wide spectrum of fields. Featuring compact volumes of 50 to 125 pages, the series covers a range of content from professional to academic.

Typical topics might include:

- A timely report of state-of-the art analytical techniques
- A bridge between new research results, as published in journal articles, and a contextual literature review
- A snapshot of a hot or emerging topic
- An in-depth case study or clinical example
- A presentation of core concepts that students must understand in order to make independent contributions

Briefs allow authors to present their ideas and readers to absorb them with minimal time investment. Briefs will be published as part of Springer's eBook collection, with millions of users worldwide. In addition, Briefs will be available for individual print and electronic purchase. Briefs are characterized by fast, global electronic dissemination, standard publishing contracts, easy-to-use manuscript preparation and formatting guidelines, and expedited production schedules. We aim for publication 8–12 weeks after acceptance. Both solicited and unsolicited manuscripts are considered for publication in this series.

More information about this series at http://www.springer.com/series/10028

Renata Rawlings-Goss

Data Science Careers, Training, and Hiring

A Comprehensive Guide to the Data Ecosystem: How to Build a Successful Data Science Career, Program, or Unit

Springer

Renata Rawlings-Goss
Georgia Institute of Technology
Atlanta, GA, USA

ISSN 2191-5768 ISSN 2191-5776 (electronic)
SpringerBriefs in Computer Science
ISBN 978-3-030-22409-7 ISBN 978-3-030-22407-3 (eBook)
https://doi.org/10.1007/978-3-030-22407-3

This Springer imprint is published by the registered company Springer Nature Switzerland AG
The registered company address is: Gewerbestrasse 11, 6330 Cham, Switzerland

To Malcolm and Marcus for whom all things are possible

Acknowledgments

I would like to deeply thank Kendra Lewis-Strickland, Illona Sheffey, Stanford Goss, and Keith Rawlings for their unwavering support in preparing this manuscript. Thank you for diligently searching through references, editing, and genuinely strengthening this work with your unique gifts.

Contents

About the Author

Renata Rawlings-Goss is a Data Strategic Coach and Author who helps professionals and executives cut through the clutter of the Data Science landscape to find their niche.

She is the current Executive Director of the South Big Data Regional Innovation Hub, one of only four federally funded Big Data Innovation Hubs for the nation, serving 16 states—Delaware through Texas—in forming Data Science partnerships between industry, academia, and government. She is also the Director of Industry Partnerships for the Institute of Data Engineering and Science at the Georgia Institute of Technology (Georgia Tech).

Previously, Dr. Rawlings-Goss worked with the White House Office of Science and Technology Policy, under President Obama, to create the National Data Science Organizers group. She also co-led the writing team for the Federal Big Data Strategic Plan, including 19 federal agencies, and was awarded an AAAS Big Data Science and Technology Policy Fellowship with the National Science Foundation in the directorate of Computer and Information Science and Engineering (CISE-OAD).

Dr. Rawlings-Goss lives in Atlanta, GA, and received her training in computational genomics, biophysics, and physics from the University of Pennsylvania, the University of Michigan, and Florida A&M University, respectively. Her research interests included data-driven analysis of African and African-American genetic expression as well as professional interests in Data Science education and workforce development in communities of color.

Abbreviations

AI	Artificial intelligence
HR	Human resources
IoT	Internet of Things
ML	Machine learning
NLP	Natural language processing

About the Book

This book is an information-packed overview of how to structure a Data Science career, a Data Science degree program, and how to hire a Data Science team, including resources and insights from the author's experience with national and international large-scale data projects as well as industry, academic and government partnerships, education, and workforce.

Outlined here are tips and insights into navigating the data ecosystem as it currently stands, including career skills, current training programs, as well as practical hiring help and resources. Also, threaded through the book is the outline of a data ecosystem, as it could ultimately emerge, and how career seekers, training programs, and hiring managers can steer their careers, degree programs, and organizations to align with the broader future of Data Science. Instead of riding the current wave, this book ultimately seeks to help professionals, programs, and organizations alike prepare a sustainable plan for growth in this ever-changing world of data.

The book is divided into three sections, the first, "Building Data Careers," addresses the perspective of a potential career seeker interested in a career in data; the second, "Building Data Programs," is from the perspective of a newly formed Data Science degree or training program; and the third, "Building Data Talent and Workforce," is from the perspective of a Data and Analytics Hiring Manager. Each is a detailed introduction to the topic with practical steps and professional recommendations.

The reason for presenting the book from different points of view is that in the fast-paced data landscape, it is helpful to each group to more thoroughly understand the desires and challenges of the other. It will, for example, help the career seekers to understand best practices for hiring managers to better position themselves for jobs. It will be invaluable for data training programs to gain the perspective of career seekers who they want to help and attract as students. Also, hiring managers will not only need data talent to hire but also workforce pipelines that can only come from partnerships with the universities, data training programs, and educational experts. The interplay gives a broader perspective from which to build.

Chapter 1
Introduction

1.1 The Data Science Opportunity

Data Science and analytics is not only improving businesses and sparking new industries, it is also improving the human condition under which every person needs food, shelter, clean water, health care, security, and education [1]. This need for transformation in global society shows that the relevance of Data Science is not only its unparalleled ubiquity and enormous scope, but its potential to improve lives and decision-making across a wide variety of areas. Organizations of all sizes and types are increasingly reliant on data as critical to their core operations. For some quick numbers, the Big Data Analytics market—just one slice of the larger Data Science and analytics market—will grow to over $187 billion within the decade [2]. Legacy businesses are being transformed by data while new businesses are being powered by data. The previously out of bounds worlds of politics, human rights, environmental advocacy, energy, education, healthcare, humanities, arts, and social good are now accessible and welcoming to big data and data scientists.

In the midst of all this, there is a huge opportunity for career development. Adding analytics skills now can be a boost in almost any field and create a niche or opportunity for employment. This opportunity is made even greater as statistics reveal a global talent gap in the workforce [3]. For this reason, data degree programs are rapidly emerging across the globe that take on different aspects of the training landscape. Companies are also looking to hire talented people that have a grasp on these new opportunities transforming our world!

R. Rawlings-Goss, *Data Science Careers, Training, and Hiring*, SpringerBriefs in Computer Science, https://doi.org/10.1007/978-3-030-22407-3_1

1.2 Data People: An Extended Metaphor

If you are picking up this book you may already know that Data Science is exploding, but the breadth and scope of its applications, as well as the potential careers and skills needed for work in the field, may still be opaque to you and rightfully so. "Data Science" has been used as a catch-all phrase encompassing different sets of skills, job descriptions, and activities. Therefore, the phrase "training to be a data scientist" can be misleading, so we approach the field of data science by comparing it to another well known field, healthcare.

The Data "Doctor" Whenever someone approaches me with the question "How do I become a data scientist?". I always ask, "What kind?" The phrase "I want to be a data scientist" is like saying "I want to be a doctor". Do you want to be a cardiologist, a surgeon, an internist, or a podiatrist? It makes a huge difference! Everybody is proud and excited that you want to be a doctor, but for YOU the salary, the day-to-day life, and the years of training will be totally different depending on your choice. The same is true of Data Science. If you say you want to be a data scientist, do you want to be a data engineer, a data visualization expert, a business analyst, a data researcher, a database manager, or a creative combination of any of these? These roles can be completely different in the salary ranges, the work required, as well as the required level of proficiency and training needed. Different data roles are a part of a bigger ecosystem of data professionals, akin to nurses, medical assistants, and specialists in healthcare.

Data Generalist Unlike healthcare, however, Data Science as a field is not well developed with clear job roles and training programs for each role. It is in its early days, where "data scientists" are by in large lumped together and asked to fill a number of niches simultaneously. This has led to the rise of the data generalist. The data generalist can be compared to the rural doctor where the rural doctor was the catch all for everything in one small town. Whatever your problem, go see "the Doctor" and they were able to fix most of your common ailments. It was a simple solution, but each town was limited in what could be treated by the scope of the doctor's knowledge. Data Science is the same. Right now, companies, organizations, and nonprofits are hiring for Data Generalists and relying on those persons to fix everything for the organization. The same limitations apply; organizations are limited by the training and expertise of the data generalist. As knowledge blossoms, therefore, the rise of the data specialist is needed to keep pace.

Data Specialists With data everywhere, the scope of the available knowledge on any one topic is outpacing the limits of what any one person can keep up with and master effectively. Consequently, since 2018, there has been a distinct trend toward specializations. Data science teams, as opposed to single analysts, are growing. It is now more likely that data scientists will be working with other data scientists, analytics professionals, and data engineers [4].

Data Care The metaphor of healthcare also helps explain the main secret of Data Science; it is about people. While healthcare could be said to be about biology, drug interactions, operating procedures or ethics policies, it is ultimately about caring for people's health with the main goal of making people healthier. Similarly, Data Science can be said to be a mix of computer science, math, statistics, and business ethics, but is ultimately about caring for people's data with the main goal of making people's lives and work more efficient and insightful. With that in mind, later in the book when we introduce the mix of skills needed for a data career this is why this list includes "innovation skills" such as problem solving and discovery listening. These innovation skills are hugely important to employers and must be taught by data training programs and taken seriously by individuals as well. Your true intent (ethics) as well as how you communicate that intent (communication/presentation skills) can win or lose you the ability to affect change in your business or career.

References

1. Business Higher Education Forum. 2016. *Data Science and Analytics Higher Education Survey.* Gallup and Business Higher Education Forum.
2. International Data Corporation. 2016. *Worldwide Semiannual Big Data and Analytics Spending Guide* . Framingham, MA: IDC.
3. Business-Higher Education Forum. 2016. *Creating a Minor in Applied Data Science.* Business-Higher Education Forum.
4. Burtchworks. 2018. *www.burtchworks.com.* 08. https://www.burtchworks.com/2018/08/27/beyond-the-unicorn-4-developing-data-scientist-career-paths/.

Chapter 2
Building Data Careers

The most appealing thing about a career in data is that you have the world open to you. Data is the modern gatekeeper, getting you into almost any arena. The challenge in building a data career is the misconception that the typical career path rolls out in a linear or stepwise way. There are a number of paths and specialties that can be taken or pursued, but they are not neatly outlined, as the Data Science field itself is new. We will discuss career options, the skills you may need for those options, pros and cons of different types of training programs from online courses to full Data Science degrees, including a list of over 460 Data Science degree programs across the country. We will touch on the job market, the ups and downs of management and location, as well as experiences from real data scientists.

> Now every company whether they sell shoes or personal services is a data and analytics company as well—Melvin Greer, Chief Data Scientist at Intel Corporation

You could have a career with data at the center of your work or a career using data as a tool in service to a larger mission or personal goal. Previously, if you did not know what you wanted to do the advice was to become a lawyer or a doctor. Why a lawyer? Because if you were interested in entertainment you could become an entertainment lawyer, if you were interested in non-profits you could become a non-profit lawyer, if you were interested in finance and corporate strategy you could become a corporate lawyer, and the list goes on. In any industry you need at least one lawyer in the room. Why a doctor? Because doctors are recession proof. Their skills are needed in any country, any nation, and can cross any border. Today, the same expansive entry point and global reach are both more than true of data scientists and analytics careers, the difference being there are a lot fewer qualified people for these roles. Therefore, a career in data does not mean only working for tech companies. Non-profits, universities, governments, as well as industries of all kinds from retail, fashion, film, finance, manufacturing, or media are all looking for data-literate or data-enabled people.

R. Rawlings-Goss, *Data Science Careers, Training, and Hiring*, SpringerBriefs in Computer Science, https://doi.org/10.1007/978-3-030-22407-3_2

As a citizen of the twenty-first century, the importance of data literacy is made manifest every day. Personal and corporate information based on big data is present in the news and online, and its effects are felt in many ways such as how you are approved for loans, fined for traffic tickets, assessed for job recruitment, targeted for product advertisements, as well as many others. Some basic levels of data literacy are imperative as you move through life in order to understand and make the best decisions for yourself. Taking the next step into a data career can therefore be personally useful as well as professionally fulfilling.

As we move through this book, we will expand the range of possibilities for data enabled careers, the fundamental skills/qualities needed to pursue these careers, and the pathways to high paying fulfilling jobs. We briefly discuss, picking your path, digging through job descriptions, exploring specialties, bolstering your fundamental skills, and entering the market as a leader. First, we will discuss two broad professional paths—the data generalist and the data specialist.

2.1 Assessing Data-Enabled Career Paths

Data science careers come in many combinations, so it is important to begin your journey with a clear idea of the different paths available to you. Ideally, the ability to envision your career over not just the next few years, but the next 5–10 years gives direction to your search. Newcomers, however, typically do not know where to start so they begin by exploring job titles to get a sense of the field. Most quickly learn that job titles in the Data Science can be deceiving, because the tasks required change. Also, when picturing a future career path, we imagine it to be a straight line—"I am a data engineer, so I want to become a senior data engineer and eventually a Chief Data Engineer"—without considering what that really means and if you will actually enjoy work and responsibilities of those roles. A good predictor of your happiness and success is whether you enjoy the tasks that you spend most of your time doing. Ask yourself:

> What will I be doing on Monday mornings as a data scientist? On Wednesday afternoons as a product engineer or CTO?

We will discuss some of the tasks associated with Data Science jobs, keeping in mind that many titles are interchangeable. There are often three paths through tech careers, first the "hardcore" developer who spends most of his or her time writing code. This is what most people think of as a data career and for which most bootcamps and online trainings are developed. The second is using data as an enhancement of another role such as a data-enabled professional. These individuals interpret the data gathered and visualize how to do their job better i.e. make and save more money, better use limited resources, or provide new insights, services, or products. The third is the strategic data manager, who manages a team of data professionals and contractors at different levels. Managers need a host of additional skills as well as broad knowledge of the data landscape. Finally, these three career paths can take on the flavor of a specialist or a generalist.

2.1.1 *The Data Generalist*

Data generalists are jacks of all trades. This means they are broadly trained, quick problems solvers and equipped to work on every stage of the analytics lifecycle (including data acquisition, transformation, and analytics). Generalists are in hot demand, particularly at organizations where there are many varied use cases and a smaller team to manage them and are among the more high-paying roles. Data generalists are largely what are considered "data scientists" because they need to be prepared for anything and require a deep base of knowledge from which to draw solutions. Job titles, however, do not usually list positions as generalists but rather a range of titles; some of the more common titles are described below.

Because data generalist is a broad term, it incorporates many sub-specialties of professionals including but not limited to [1]:

- Data Scientists
- Data Analysts
- Data Architects
- Data Engineers
- Database Administrators
- Business Analysts
- Data and Analytics Managers

The common thread is that generalists are expected to have strong and broad technical skills (mathematical techniques, comprehension of statistics, and technology-based skills) as well as specific non-technical skills (interpersonal skills, superior communication abilities, and strong teamwork capabilities). Both of these skill sets are discussed in more detail in the section on Fundamental Skills.

Each of these varied roles and job titles require a unique emphasis on skills from an array of academic specialties including math, science, statistics, and technology arenas. The professionals within each subset of the Data Science field provide their own unique perspective on the methodology used to take their vast array of skills and turn them into working solutions for small businesses and large corporations alike. We will talk briefly about the common distinction between roles, keeping in mind that there are no standards and different places define their roles differently [2].

The Data Scientist (Problem Solvers)

Data scientists, as the name suggests, are truly scientists or problem solvers because they use a high paced scientific method to diagnose a problem, determine the method and tools needed to solve it, explore what works, and try again. This fusion of curiosity to explore current trending problems, combined with the ability to use data to create solutions to these complex issues, makes a data scientist a powerful tool in predicting the future of an organization. Part mathematician and part computer scientist, data scientists are usually experts in gathering data on a problem, interpreting that data, and applying or building tools to apply that knowledge into action.

The Data Analyst (Translators)
Every business, no matter their specialty, collects information on an ongoing basis about its operational costs and profits. This data must be analyzed and studied, using various techniques, then applied to improve the functionality of a company and enhance its profitability. This is where a data analyst comes in. Data analysts translate statistical information into a language everyday people can understand and use as a basis to make practical decisions. Data analysis might be used to help source more affordable materials, to reduce transportation costs involved in operating a business, or to track down issues that are costing the company too much money.

The Data Architect (Builders)
As a result of larger and larger slices of business sales being conducted online, new websites are offering innovative products and services daily. All of these websites require builders and extensive data analysis to function with a successful profit margin to ensure their data systems are not overwhelmed. Like a building architect, a data architect designs the structure needed to house data and ensure it is effectively gathered, stored, and managed. Data Architects design the infrastructure needed to help businesses make a solid name for themselves in a seemingly endless sea of competitors by use their analytical skills to build tools to interpret the information needed to help digital businesses succeed.

The Data Engineer (Testers)
Data engineers test, evaluate, and improve on the solutions designed by data architects. Companies are increasingly looking for professionals who can help them facilitate a secure and smooth movement of data across all networks. Data engineers have a precise knowledge and understanding of software engineering and testing patterns, combined with coding experience, in order to create robust and usable solutions. Data engineers build on what data architects design, by developing and maintaining the infrastructure that keeps data secure and moving smoothly [3].

Database Administrator (Librarians or Archivists)
In this information and technology age, gone are the days when critical data was stored in filing cabinets on paper. Nearly all vital information that a company relies upon to improve its bottom line is stored in computer databases. Database administrators are responsible for the important task of storing and backing up this information in physical and virtual spaces. In case something goes wrong with the original software or the manner in which it is stored, database administrators ensure that this data is easily accessible to those who need it but simultaneously is kept safe from unauthorized access. They essentially serve as the gatekeepers and protectors of the vital information that helps keep businesses operating successfully.

Business Analyst (Communicators)
Business analysts are generally a little less technologically savvy but offer a deep knowledge of the processes involved in running a successful business. They are masters at connecting this insight to real world strategies that enable businesses to become more successful. Consider them the intermediary that merges the technical

side of an organization with the business mission. They bring the two together to offer solution-oriented strategies targeting successful business operations.

Data and Analytics Manager (Coaches)
Data and analytics managers are not only the coaches but often the cheerleaders of the team. They are responsible for ensuring that the right people are hired as well as the right goals and priorities are set for all parties. Data and analytics managers require strong social skills of a different type than senior engineers, analysts, or scientists. They need to lead a team of independent individuals, almost like a chair of a committee or department, and also have the technical know-how to analyze and validate data findings. Running a successful data team and overcoming obstacles is a complex series of events that requires professional input from a variety of specialists. Data and analytics managers, therefore, must cope with all of these backgrounds, provide cohesion, and offer a wide variety of career choices that are exciting and solution-oriented for those who are technically inclined.

Other job titles such as Statisticians, Mathematicians, Computer Scientists and Programmers of all kinds, provide the very backbone for the field of Data Science, and are also titles that may be playing the role of the data generalists within an organization. There is room for all manner of professionals that offer the experience and expertise needed to gather and analyze data and use it to come up with real solutions to problems.

2.2 The Rise of the Data Specialists

> Don't become an expert in a tool only; be a lifelong learner for adaptability and career resilience—Renata Rawlings-Goss

While companies that are hiring their first data scientists may still look for data generalists who can wear every hat, many organizations are now looking for data specialists to fill out teams, where each member of the team has a more specialized background. The trend towards specialization is increasing as companies are able to really focus on the specific use cases for Data Science that are relevant to their business. I will explain two specializations that have seen an uptick in positions, specifically *Artificial Intelligence*, which includes Machine Learning, Computer Vision and Imaging, and Natural Language Processing (NLP), and the *Internet of Things (IoT)* [4].

2.2.1 Artificial Intelligence: Machines Acting Like People

One popular topic is Artificial Intelligence (AI), either from professionals and students looking to get into careers, companies looking to leverage it in their business, agencies and nonprofits trying to improve their missions, or smart cities trying to understand their citizens and build out requirements. While AI is overwhelming the conversation right now, I find that some very smart people in the categories above do not really have a complete grasp of what AI is. So, I thought it might be helpful to lay out how I begin these conversations by simply explaining what AI is, along with three of its most popular career specialties.

The words *Artificial Intelligence* seem to conjure up movie images for most people, but the meaning of AI can be simple.

> Artificial Intelligence is anything that helps machines perform tasks that are characteristic of human intelligence, such as learning from the past, understanding language, and seeing and recognizing objects and images.

The dream of movies like The Terminator, The Matrix, and Ex Machina is to get AI "perfectly packaged" into "human-like robots" with all of these elements working perfectly together. The real uses and meaning of AI, however, are a little more commonplace and broken up into separate fields, but no less interesting.

Artificial Intelligence can be broken down into individual "human-like" tasks, like the ones below. Each human task listed is a separate field with its own body of knowledge (Fig. 2.1).

Fig. 2.1 Artificial Intelligence Sub-fields and how they relate to human intelligence. Machine Learning (ML) is an attempt to mimic the task of learning, Computer Vision (CV) the task of recognizing objects, Natural language Processing (NLP) the task of understanding language and the Internet of things (IoT) the task of experiencing the physical world. Image was provided under a creative commons license with text modification. Original from WOC in Tech Chat Stock Images @ Microsoft NYC

1. Learning from the Past is Machine Learning (ML)
2. Seeing and Recognizing objects and images is Computer Vision (CV).
3. Understanding language is Natural Language Processing (NLP),
4. Experience the physical world is the Internet of Things (IoT)

Of course, there are many other applications of Artificial Intelligence such as recognizing speech, planning, or robotics, however, these are the four where I have seen an increased demand for workforce development. Although there is only a very small group of people who are experts in all areas of AI, the good news is you do not have to be. Specializing in just one area puts you in an elite status. I will describe a little more about the above three areas: common use cases, what you need to learn to get into these fields as well as who is using/hiring people for this work.

2.2.2 *Machine Learning (ML): Learning from the Past*

Machine Learning is the application of AI that allows computers to learn, but only in a specific way—pattern recognition. These patterns are found within Data.

Machine = Your machine or computer
Learning = Finding patterns in data

The heart of ML is "training" a machine to predict what will happen in the future by providing a lot of examples of what happened in the past. The training process requires providing huge amounts of data to a machine and allowing the machine to adjust itself to improve its future predictions. As opposed to hard coding millions of lines of software routines with specific instructions, this allows patterns to be detected from the past that can help predict what might happen in a similar, but not identical, situation in the future.

Common Uses

- *Social Networks*: Twitter Curated Timelines, Facebook chatbots, Pinterest Content Discovery, Match.com Predicting certain match preferences for better compatibility.
- *Finance and E-commerce*: Equifax predicting fraudulent activity on a credit card, United Healthcare predicting fraud waste and abuse in insurance claims. Multiple companies: Predicting customer conversion rates or improving customer experience and segmenting customers for better targeted services.
- *Health and Biology*: Predicting patient diagnostics for doctors to review, finding patterns in gene mutations that could represent a precursor to cancer.
- *New and Cool*: Academic institutions and companies like Google have been very busy in recent years, having diversified ML into such fields as anti-aging technology, medical devices, environmental monitoring, and—perhaps some things that range on the bazaar such as—machines that "dream" [5].

What to Learn: (1) R or Python, (2) Basic descriptive and inferential statistics languages [6], (3) Data Exploration/Cleaning/Preparation (4) Introduction to Machine Learning (Supervised learning, Unsupervised learning and Reinforcement learning), (5) Advanced Machine Learning (Deep Learning, Ensemble Modeling, and Machine Learning with Big Data) (6) How to Participate in the world (Competitions, Real-world projects, and understanding developer ethics and impacts of ML on today's society).

Companies Using Machine Learning: Facebook, Salesforce, Twitter, IBM, Microsoft, HubSpot, Google, Stanley Black & Decker, and many others.

2.2.3 Computer Vision and Image Processing: Recognizing Objects and Images

Computer vision and image processing is also a component of artificial intelligence that allows computers to process images closer to the level at which the human brain functions. It involves AI systems that are tasked with analyzing visual information from static images as well as videos. Every day, more opportunities to automate and assist are making themselves available for an insurance agent, a doctor, a beauty consultant, or even the driver of a car.

Common Uses

- *Car Crash Reports*: Computer vision specialists are creating systems to look at car accident photos for insurance companies to accept claims [7].
- *Medical Images*: Medical imagery such as X-rays, MRI scans, cancer screenings, and biopsies are being looked at for hospitals to diagnose diseases [8].
- *Fashion and Beauty*: Customer's faces are photographed for beauty companies to provide better product suggestions [9].
- *Self-driving cars*: Self driving cars rely on advanced image processing to navigate the world around them.
- *Social Media*: Facebook tagging suggestion feature.

What to Learn: Deep Learning, Machine Learning, GPU/CPU Programming, Object Detection, 3D Reconstruction, Pattern Recognition, C++, OpenCV, Matlab, Python, Spark.

Companies Using Computer Vision [10, 11]: Gentex Corporation, A9, COGNEX, HP Labs, Philips Healthcare, Interphase, NOKIA, KLA-Tencor, Immersive Labs, Intuitive Surgical, IBG, Imimtek, PPT Vision, HHMI, Occipital, Northrop Grumman, Justin.tv, Incogna, Video Surf, tyzx, Object Video, Honeywell, SkyBox, Park Assist, AiLive, Hover, Aptina, OptraScan, Honda Research, Image Metrics, Pelican Imaging, Graftek Imaging, NVIDIA, Microscan, DAQRI, Kairos, LTU Technologies, Code Laboratories, FotoNation, Imerit, Sighthound, Second Spectrum, GE Global Research, Yahoo!, Optra Systems, Vizzitec, Pixuate, and others.

2.2.4 *Natural Language Processing (NLP): Understanding Written Language*

Natural Language Processing (NLP) is an application of artificial intelligence aimed at allowing computers to understand the written word. It has the ability to allow computers to automatically extract insights from text, such as documents, emails, videos, tweets, and other unstructured material, as well as analyze and synthesize company reports or publications. The simplest form of NLP is to reduce every document to a "bag of words" and allow a user to search for a "keyword" in all of the assembled documents.

Common Uses

- *Search Engines*: Clearly, sophisticated search engines like Google have taken NLP far beyond simple keywords. It can be used as the foundation of private search engines to allow for keyword search through a large number of internal or external documents.
- *Sentiment Analysis*: Categorizing opinions from text and is commonly used to examine customer reviews or tweets.
- *Healthcare*: With lengthy patient records and doctors' notes, NLP can be used to evaluate insurance claims or to better understand patient outcomes.
- *Call centers*: NLP is also put to use in call centers to more efficiently serve customers or better understand user needs.
- *Speech Recognition*: NLP is used as a part of a number of technologies. Virtual assistants, like Siri or Alexa, translate the voice data into written words then uses NLP techniques to extract meaning from what was said.

What to Learn: The two most common software frameworks used in NLP are Solr and Elasticsearch. Both are open source, meaning they are free to download. There are many tutorials as well as conferences that could be attended to learn more. Both operate similarly and have comparable features at this point thanks to a worldwide developer community building new features for each. Different companies and organizations use one or the other so if you have a specific employer in mind check to see what platform they are using and learn that software. *Additional skills/tools*: Python, Machine Learning, Deep Learning, TensorFlow, Neural Networks, Computer Science, and Computational Linguistics.

Companies Using NLP [12–14]: Apple, American Express, Wells Fargo, Staples, AT&T, Allstate, United Healthcare, FDA, Monsanto, Uber, Reddit, RedHat, eBay, E-trade, CareerBuilder, Intuit, Cisco, AOL, City search, Chegg, Kickstarter, TaskRabbit, Instacart, and many others.

2.2.5 *Internet of Things (IoT): Sensors Bring the Internet into the Real World*

The use of sensors (and the data they produce) has exploded over the past several years. The Internet-of-Things (IoT) is the process of attaching sensors to physical objects, like chairs or lights, and connecting those sensors to the internet. The relationship between AI and IoT has been described as akin to the relationship between the human brain (AI) and the body (IoT). All of the connected sensors that make up the Internet of Things are like our bodies, they provide the raw data of what's going on in the world. Artificial intelligence is like our brain, making sense of that data and deciding what actions to perform [15]. Internet-of-Things data practitioners can be employed for a wide range of projects.

Common Uses

- *Maintenance Prediction*: Using advanced infrared thermal imaging, vibration analysis, and other sensors to predict when vital machines will fail or need maintenance for industries like manufacturing and transportation [16, 17].
- *Emergency Rooms*: ER's exploring the use of sensors to optimize bed allocation and patient stays for different conditions to effectively provide care.
- *Environmental Monitoring*: Monitoring air quality along major roadways, near hospitals or schools.
- *Traffic Monitoring*: Tracking car volume and near miss crashes to optimize city planning efforts.
- *Inventory Tracking*: Companies using sensors to more effectively measure inventory and manage their supply chains [18].

What to Learn: Streaming Data Analysis, Python, Machine Learning, Sensor Data Systems, and Telematics.

Companies Using IoT: Mattel, Kohler, Honeywell, Microsoft, Emerson, Geha, Electrolux, Randall and Reilly, Phillips, and many others [19].

2.3 Skills for the Career You Want

> The true goal is the extraction of value from data, which starts in the mind. Data has no intrinsic value, only what can be imagined and made real.—Renata Rawlings-Goss

As you can see from the section above, different skills are required for different data careers. As data creeps more and more into not only our work but our lives, it becomes an increasing fundamental skill, like literacy or numeracy, in order to be an active citizen of the world. There are some basic technical skills that span several well-defined disciplines, specifically statistics, computer science, and math, as well as basic programming to manipulate, clean and extract data which are fundamental. Also, there are some basic non-technical skills I call "Innovation skills" that are equally vital to success.

2.3.1 Innovation Skills (Not "Soft Skills")

I dislike the term "soft skills". Some use the term "soft skills" to refer to skills like communication, curiosity, work ethic, etc. This does a disservice to the importance of these skills in the life of a data professional. It makes you think of soft throw pillows, decorative but not necessary. When in fact they are critical skills needed for data career success, of which we will discuss concrete examples.

Data science requires an innovation mindset that is employed consistently in order to be valuable. Innovation comes from the transformation of an industry or process and there are four tangible steps needed before any technical skills get employed, to avoid wasted time, money, IP, and opportunity.

2.3.2 Four Invaluable Skills for Success

Converting Wishes into Problem Statements We all think we are good problem solvers, but what does it really take to problem solve with data? Primarily, it takes a clear understanding of the problem! This is a non-trivial step often ignored and assumed. Most companies and individuals come up with a problem they are trying to solve, for instance, "We want to increase the effectiveness of our call centers. Please use data to solve this." Which sounds like a fully defined problem, but the moment you scratch the surface you see that there is so much we do not yet understand about the problem.

The above example is not a problem statement at all. It is a wish. People and organizations pass on their wishes to data analysts and it is often the data professionals' job to turn them into problem statements. For this example, the analyst would first need to know how the call center is currently operated down to all the nuts and bolts of how calls are processed. Are they all recorded or only a subset? What do people call about most often? Where do they call from? Who are the most productive operators/agents? How is the data stored? What is the normal flow through for the customer experience. This background information gathering is critical to understand the situation which will be different at each organization. Consider two examples of good problem scoping from the Center of Data Science and Public Policy at U. Chicago [20]:

Example 1: Lead Poisoning The Chicago Department of Public Health was interested in preventing lead poisoning in children. The initial goal was to increase the effectiveness of their lead hazard inspections. One way to achieve that goal would be to focus on homes that have lead hazards. Although helpful, this approach wouldn't get to their real goal, which was to prevent children from getting lead poisoning. Finding a home with lead hazards and getting it remediated is only beneficial if there is a high chance of a child present (currently or in the future) who is likely to get exposed to lead. The next iteration of the goal was to maximize the number of inspections that find lead hazards in homes where there is an at-risk child (before the child gets exposed to lead). Eventually, the final problem statement was found: identifying which children are at high risk of lead poisoning in the future and then target interventions at the homes of those children.

Example 2: High School Graduation One of the bigger challenge's schools are facing today is helping their students graduate (on time). Graduation rates in the US are ~65%. Schools are all interested in identifying students who are at risk for not graduating on time. When initially talking to most school districts, they start with a very narrow goal of predicting which kids are unlikely to graduate on time. The first step is to go back to the goal of increasing graduation rates and asking if there is a specific subset of at-risk students they want to identify? What if we could identify students who are only 5% likely to be at-risk versus students who are 95% likely to not graduate on time without extra support? If the goal is just to increase graduation rates, the first group is (probably) easier to intervene with and influence while the second group may be more challenging due to the resources they need. Is the goal to maximize the average/mean/median probability of graduating for a class/school or is the goal to focus on the kids most at risk and maximize the probability of graduation of the bottom 10% of the students? Or is the goal to create more equity and decrease the difference in the on-time graduation probability between the top quartile and the bottom quartile? All of these are reasonable goals, but the schools have to understand, evaluate, and decide on which goals they care about. This conversation often makes them think harder about analytically defining what their organizational goals are as well as the tradeoffs.

Both examples show the power of problem scoping. The first example produced a problem statement that is analytically actionable, where a prediction could take place based on past events. The second example highlights how a broad goal can be analytically interpreted in many ways and that the scoping process alone highlights key decisions that need to be made before work should begin.

Listening to Domain Experts The information gathering process needed for problem scoping leads directly into the second innovation skill, listening. Often the best or only way to get background information is to talk to a number of different people, get their buy-in, and ask their opinions. Technical people or senior management assume, most times wrongly, that they know how things are working, but no matter what the managers say about how it should be working there are always differences at ground level. This is a very tricky proposition for the average technically inclined person. Unless there is an intermediary such as a Data Science manager or translator, this will fall on the analyst. Even if there is such a person, the game of telephone between parties provides messages with meanings that do not translate perfectly. Being able to overcome this barrier of needing to code alone, being the lone data person dreaming up a solution, or going away for 2-weeks and coming back with a product, will accelerate your career by leap and bounds. For one, you will not waste time on beautiful solutions nobody wants or will use.

Listening skills also help identify places in the pipeline that could be tightened up as well as solutions that people will actually use. The perfect solution can be built, but if people deem it too cumbersome, authoritarian, or complicated they will not use it, or they will subvert it in some subtle way. Case in point, the CDC came out with a report that one in six truck drivers do not wear seat belts and that 40% of the fatal accidents could have been prevented if they buckled up [21]. So, an enterprising executive decided to use IoT sensors to ensure their drivers were buckling up. The solution was a sensor on all of the seat belts that could tell if it was clicked while in motion and alert the controller if a driver was delinquent. It was simple and elegant and seat belt use numbers went way up. However, one reporter found that drivers, particularly those who had to get in and out of the trucks a lot, were just clicking the seatbelt behind them and sitting on top of it to allow for a quick exit without getting dinged by their supervisors.

This was a case in which the data and monitoring were not enough to change behavior. Here it would have been critical to ask, "does my solution require a significant behavior change in the company, workers, executives, or business units?" If so, any such solution would need buy-in first, which means you will need to be able to explain how it benefits everyone, without being confusing.

Communicating and Presenting Benefit A common vision of data careers pictures a tech genius holed up in a room alone or with only other tech people coding away at brilliant solutions. I have seen this happen, but also, I have seen truly brilliant solutions sit idle or get shelved completely by not understanding this one fact. People want to understand what you do in plain language, be included, and feel they could communicate the benefit. This includes how you present or visualize your data, how you talk about not only your data but the entire story of how it fits into the original question.

This is important not only to get buy-in before a major project is started, but also in explaining outcomes, results, and demonstrating a return on investment. Data science professionals need to be able to accurately convey the results of their work to both technically inclined individuals as well as those who are not tech savvy. To accomplish this, their interpersonal and communication skills need to be outstanding.

Curiosity and Continuous Learning Now more than ever it will be *crucial* for analytics professionals and data scientists to keep their skills sharp and current. This means a commitment both from the individual, to invest their personal time, but also from the organization to foster and support training (and retraining) as part of doing business. In a market where your skills can become obsolete in just a few years, staying curious is the best way to make sure you keep your skills fresh and marketable.

Even for top-level management positions, it has become clear that staying hands-on with data projects is key. If you are looking to be a senior-level data scientist in a management position, it is more important than ever to stay current. Many companies are looking for leaders who can be what is known as a "players coach", in that they can be hands-on enough to mentor the team as well as lead. This is especially true as Data Science expands rapidly into new industries and leaders will often be the first hire to a team. With the skills gap still presenting a significant hurdle for many companies, more leaders will need to look to train their current employees to fill their analytics needs. This in-house talent has the added appeal of domain experience but may require leaders to find or provide additional experiences.

Data science professionals need to maintain a degree of curiosity to be able to detect current trends in their field and use them to make future predictions based on the data they collect and interpret. This natural curiosity will inspire them to stay on top of their game in terms of continuing their education as well.

2.3.3 *Technical Skills*

Once the Innovation skills above are mastered there are a few basic technical skills that will be invaluable to the data professional.

Matching Tools to Problems Once a specific area of improvement is identified and people have bought in, the fundamentally technical parts begin starting with the simple question of what are the best tools for the job, a non-trivial question. Matching the correct tool or set of tools to your problem requires an overall knowledge of the uses and limitations of various techniques as well as whether the organization has the staffing, resources, or resolve to pursue every path.

Correcting Dirty Data While it's true that there are some programs that claim to take care of this issue, in many cases the ability to locate and correct corrupt or imperfect data is one of the most important and time-consuming skills in the field of Data Science. Some surveys of data scientists say they can spend up to 80% of their time on this task [22]. This skill is highly important among smaller companies, where incorrect data can so vastly impact their bottom line. This set of skills can include locating and replacing missing values, correcting inconsistencies in formatting, and changing timestamps.

A Good Understanding of Basic Statistics, Calculus and Algebra In order to understand and accurately use most Data Science tools you will need some fundamental statistics skills. For example, data scientists need to be familiar with maximum likelihood estimators, statistical tests, distributions, calculus, and algebra. It may seem odd that a Data Science professional would need to know how to perform calculus and algebra since many of the programs and software used today can do all of that and then some. However, newer businesses, whose products are defined by data may require small continuous advancements to stay competitive. They will favor professionals who have these skills and don't rely on software alone to get the job done. Note: The full list of skills needed will be highly dependent on the industry chosen. It is vital that professionals know how to recognize which technique is going to work best in a given situation which requires a thorough understanding of its limitations.

Data Visualization Skills This skill subset is of particular importance to companies that are just learning to make decisions based on data. Narrative Visualizations are an accessible way to tell the story of the data and lay a foundation from which future predictions can be employed.

2.4 Learning Data Science

> Data science education is currently offered in many forms, and for the near future this variability is expected to continue due to the rapidly changing nature of the field. There are however some common delivery methods that have their pros and cons—The National Academy of Science [23]

Data Science trainings are released at an ever-increasing rate. This book contains a listing of over 460 Data Science degree programs in the US and EU alone (See Resources). There are also hundreds of bootcamps and thousands of in-person and online short courses. How does one choose the best option?

After choosing a data-enabled career path and identifying the skills you need for it, you must acquire those skills. The key is acquiring skills in a way that is best for you and most recognized by the job you are seeking.

The most common forms of training, in order of time commitment, are listed below:

- A Major in Data Science or a related field
- Two-year Degrees or Certificate
- A Data Science Minor or track in Data Science
- Learning Oriented Employment
- Bootcamps
- Open Online courses
- Free Online Videos

Within these categories there is a huge variation in the method, style, and focus on what is taught. By in large, however, each has its best use case.

2.4.1 Evaluating and Finding the Perfect Pathway

Data Science Majors or Related Degrees (Best Use: To gain advanced skills, get direct career services, and fully leverage the name recognition of a university brand and degree) Due to uncertainty in the academic landscape and the rapidly changing nature of Data Science, there is no standard definition of a Data Science

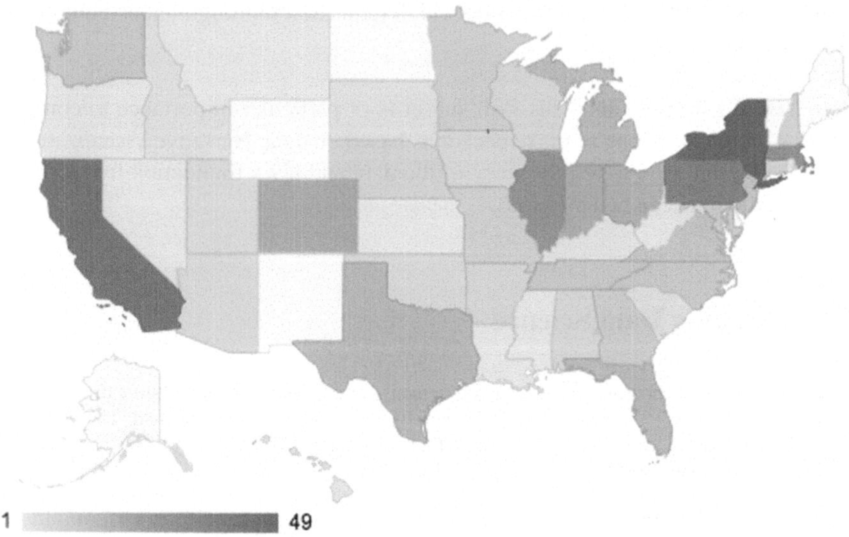

Fig. 2.2 Heatmap of U.S. data science programs by state

curriculum. There are, however, many emerging programs including over 460 analytic related post-secondary programs in the United States (See Table 6.1) (Fig. 2.2).

Of that number approximately, 300 are Masters programs, 100 are Certificate programs, 40 are Bachelors programs, 20 are Doctoral programs, and seven are Associates programs (See Table 6.1). Most masters programs are housed either partially or completely within the business school, as most jobs are going to be business enabled.

Many data or analytics programs have partnered with the private sector. Some accepting industry proposals for multidisciplinary practicums offered as capstone courses that help students gain experience and help companies address their business problems, for example at the Georgia Institute of Technology. Some offer students up to 12 months of pre-graduation work experience through co-ops and internships, such as at Northeastern University. Some universities efforts are directed at using Data Science and analytics for social good and economic development such as reducing infant mortality, accelerating drug discovery to fight disease, and realizing autonomous systems for transportation and agriculture, for example at North Carolina State University. To evaluate a Data Science degree program you are interested in attending, (See Chap. 3) for a comprehensive guide to recommendations for degree programs. It will give you a sense of the elements expected in a top program and give you the perspective to ask questions of programs of interest. Time commitment: Average 4 years.

Two-Year Degrees (Best Use: Technical colleges and community colleges are useful in getting a good foundation in Data Science tools while leaving the option open to transfer to a four-year institution.) DJ Patil, the first United States Chief Data Scientist, credits his community college education for the gifts of confidence, the ability to write, and love of mathematics that led him to an advanced degree and productive work in industry and government. Community colleges and technical colleges are well positioned to be highly effective providers of Data Science education while also serving as important partners for 4-year institutions that are considering the emerging role of Data Science education [25]. They create mechanisms by which students can certify specific or general skill sets with certificates or associate degrees; build foundational, translational, ethical, and professional skills to support matriculation into 4-year college Data Science programs; and provide opportunities for advanced high school students to begin Data Science training early. These institutions, and other private or for-profit credentialing providers, also serve a key role for low-income, minority, and first-time college students. Time commitment: Average 2 years.

Data Science Minors or Certificates (Best Use: Supplementing an existing degree, where intermediate Data Science skills are connected directly to a major field of study.) Data science minors are springing up at a number of universities in all areas of the country, like the University of Pennsylvania, Stanford, Rice, Berkeley, University of Michigan, Northeastern University, Iowa State and others. As such, if you are currently in school, they are a good way of determining if your interest in Data Science is a response to the possible salary bump or a true desire for the field. Also, if you are not yet sure exactly what you'll need to learn, then a minor in Data Science is likely the best path to take. It's a less expensive learning trajectory when compared to fully committing to a higher academic degree. Additionally, it will give you the opportunity to see if your desire to be a data scientist can withstand the required number of math, statistics, programming, and technology-intensive coursework coming your way.

If you are not entirely sure that even minoring in Data Science is right for you at this point, then a certificate may be an option to try. The emphasis here is the word *try*. Why? Data science certificates aren't designed to give you a comprehensive education throughout each of the Data Science domains: math, statistics, Data Science tools and processes, and programming techniques. Rather they are designed to teach how these domains specifically apply to all things data. For example, Harvard offers a Data Science Certificate [26] via their Extension Program and while it is pricey, you only need to complete a statistics course, two electives, and a single Data Science course. Even though having anything Harvard related to schooling can boost your status in the eyes of an employer, five Data Science courses are the academic equivalent of a snack—it's a small glimpse of Data Science. These are not degree programs, but they can provide you with immediate exposure to Data Science and the brand of powerful institutions to bolster your resume or CV. Time commitment: Varies.

Learning-Oriented Employment (Best use: For students or those not currently employed full-time. Those interested in dabbling in several industries or those with flexible schedules who know they are looking for a career change.) Learning Oriented Employment i.e. Internships /Fellowships/Apprenticeships or on-the-job training. Best used if you already have some fundamental Data Science skills such as a degree in statistics, math, or computer science, but do not have an integrated Data Science background and would like to test out different industries. This is a very targeted approach in which you directly interact with your potential dream employer. Recommendation: Excelling in this format is the most direct link to a full-time position provided the organization is hiring for the role you are in. From the company's' perspective, they are interacting with top talent. The downside is that it can be very industry specific so doing internships at more than four companies gets unwieldy and very time consuming unless you are also getting a degree. It is also not ideal for working professionals who do not have flexible schedules, as taking the time for a fellowship would take away for their day to day duties. For working professionals, a job where there is significant support for training and upskilling can be the best of both worlds. You can get the training you need while still working. There are only a limited number of companies, however, that have reached the maturity to design extensive in-house Data Science training, but many offer incentives to get external on the job training. For example, Booz Allen Hamilton has extensive training for employees including Mentoring Learning Circles, Data Science bootcamps, Hack-a-thons, Tech tank programs, and Kaggle style competitions, where companies like GE offer education reimbursement or incentives to employees. Time commitment: Varies.

Bootcamps (Best use: To learn a particular set of techniques directly connected to job prospects i.e. DataCamp, Flatiron School, Data Science DoJo or General Assembly.)

Bootcamps are another example of alternative credentialing, or as it is known by workforce professionals, "customized training" [24]. Better ones are customized to a particular type of skill. I caution you to be aware of two important questions before signing up for a bootcamp: (1) Does it teach the type of skills for the jobs I am aiming for and (2) Do they have direct connections to those employers? Here is why. A bootcamp is an alternative credential and will not be recognized everywhere as proof of skills. Some companies are very comfortable hiring people with bootcamp training only, but these are primarily the larger type that will require extensive interview testing and offer on the job (OJT) programs. The best way to know if a company trusts and respects the bootcamp you are exploring is to see if they have any formal partnership with the program. Many seasoned programs look to industry affiliates to help them design content that will be relevant to the market. The proof is in the hiring, so ask if their industry affiliates hire the bootcamp graduates. Another good indicator is if the company sends a lot of its current employees to the bootcamp to upskill. If so, the company sees the training as valuable and is paying for its employees to take it. Recommendation: Bootcamps are very specialized deep dives, so they would not be a good choice for a generalist. Also, make sure the bootcamp you want to take is recognized and appreciated by a company you want to work for, preferably companies have a formal relationship with the bootcamp to ensure your time and money translates into job prospects. Time Commitment: Average 1–10 weeks.

Open Online Courses (Best use: To get a broad introduction to a field or topic area.) Massive Open Online Course or paid courses, as offered by platforms like Coursera, EDX, or Udacity, focus, by in large, on an introductory level. They attract beginners looking for credentialing or to get to a relatively intermediate level in a short span of time. Credentialing is a viable option for industry professionals to enhance or learn new skills related to Data Science. There are several excellent Data Science introductory courses offered by top universities through these platforms. Recommendation: Use Online Courses as a low-cost way to get a broad overview of an area. The content is connected and sometimes you can get college credit if you complete the course and pay a fee. An added perk is that the instructors are sometimes live, and you can submit questions if you keep up with the course in real-time. Note ∗ In order to appeal to a massive audience, courses can be hit or miss with providing you the right practical skills. Some tend to focus on theory but give a solid foundation for learning other skills. Time Commitment: Self-paced or Real-time.

Free Online Videos (Best Use: When you are learning on the job and just need one or two additional skills for a particular project.) Free online videos, such as those found on YouTube, vary in quality and do not carry with them any certification of competency to get a job. Also, they have not been designed as complete programs so could be missing critical pieces. With that said, they are a quick, easy, and inexpensive ways of gaining practical knowledge on a technique. This is particularly true of new techniques that might not have made it into courses or formal programs yet. Recommendations: Use online videos to quickly gain a very specific skill to do an immediate job. Also, if learning online, watch a number of videos on the same topic for consistency (~3 to 5) and test the knowledge out on your problem right away. Note ∗ If the skill you a trying to get is so new that there are not a good number of videos on the topic, thoroughly check the author of the video. If you can find one from someone with good credentials that would be best. Time Commitment: Self-paced.

2.4.2 Picking a Career Path

Identifying a career path is the biggest step, and it is invaluable. To set a path you must assess your passions, skills, current experience, and the broader market. The intersection of these four areas will determine a solid path forward and be the basis for selecting a program or method of learning that will best suit your goals. Answering some of the following questions can clarify your thinking.

1. What activities give me energy?
2. What are my current strengths?
3. What is my current experience?
4. Can I think of a career path that incorporates the things above?
5. What skills am I missing to pursue this career path?

Picking a Program

6. Do I prefer formal instruction either online or in-person?
7. Do I want to learn on my own without a formal instructor?
8. Do I want to be able to ask questions in real time?
9. Am I looking for immediate job placement or more skills development?
10. Am I currently working and just want to supplement my skills for specific job advancement?
11. Do I want to be a technical implementer or a manager of tech professionals?

If the answer to question six is yes, and you do want to pursue a formal degree program in Data Science, become aware of interesting opportunities and career pathways before entering the university space. Identify programs/institutions that have a commitment to Data Science and analytics, whether through a center, course offerings or curricular experiences. Curricular experiences that provide the opportunity to focus on real-world problems are best, as they will offer engagement in purpose-driven, "hand-on, minds-on" work for experiential learning. Experiential learning opportunities are critical to gain the necessary skills and years of experience needed to be employable. For example, cybersecurity roles usually require a bachelor's degree and at least 3 years of relevant work experience [27]. Overall, 39% of Data Scientists and advanced analysts require a master's or Ph.D [28]. We go deeper into the market trends in the next section.

2.5 The Job Market

There is a massive Data Science job market. With jobs offered in most every sector, we will take a look at the Data Science job market in two ways, first by the numbers, then by experiences.

2.5.1 Size and Scale

Annual Data Science and analytics job openings are projected to rise steadily to 2.72 million postings per year by 2020 [29]. By 2026, overall job growth is expected to outstrip growth during the previous decade, creating 11.5 M jobs, according to the U.S. Bureau of Labor Statistics [30]. Data scientist roles have grown over 650% since 2012 [31] and job postings asking for Data Science and analytics skills outnumbered those asking for registered nurses and truck drivers combined, two of the largest hiring occupations in the US [32].

2.5.2 Skills in Demand

Of these jobs, the majority, 67%, were analytics-enabled or data-enabled jobs requiring only beginning to intermediate levels of data literacy such as familiarity with analytics methods or the ability to apply data driven techniques to problems [33]. Only 23% were true data scientist jobs which required advanced unstructured data exploration, creation of new methods, or advanced analytics. This means the market is wide open for a range of technical levels, but currently only 35,000 people in the US have Data Science skills, while hundreds of companies are hiring for those roles [34].

What this means is that there is a large volume of unfilled jobs, but most of them do not require advanced Data Science technical skills. Innovation skills with data, as described in the section on non-technical skills, are at a higher premium. Even though true data scientists can usually command higher salaries, only a small percentage of the jobs require those skills.

This is good because it means most people will not have to spend the time to become a deep Data Science expert to get into this field. Data-enabled professionals could focus on recurring techniques for a particular industry and hone their ability to apply them to different problems. This data-enabled majority also includes Analytics Managers and Team Leaders that must understand trends and techniques as well as team management and business cases [35]. These jobs are higher paying, but they also require higher levels of preparation and above-average social skills, analytical skills, or both.

2.5.3 Data Career Experiences

Beyond envisioning your initial Data Science job, think toward building a career. Some decisions have additional considerations. Let's take some case studies, I have changed the interviewees and company names for confidentiality, but the situations represent real life individuals and companies.

Angela Washington worked her way up to an authoritative role at an organization that was one the largest, best-funded, and best-positioned of its kind. The CEO was one of the biggest players in the industry. She made it onto the executive committee and ended up taking leadership of all technology issues in the organization. She learned a lot and had an uptick in opportunities that have been useful to her in her later career. The move, however, took her farther from where she wanted to be, career-wise, because it wasn't a promotion. It was a career change. She wrote less and less code, and writing code was something she genuinely loved doing. She had to turn an ever-larger number of design decisions over to others because she didn't have time to work on them herself.

Ups and Downs of Management

Understand that management can be a step up but also a step sideways. The importance of this is striking as some companies' only means of promotion is to make you a manager or executive. This is not just simply a career upgrade but really a career change as the skills needed become increasingly different. Be sure the promotion track gets you to where you are trying go and not further away from what you love. In most organizations, there's a hierarchy of managers but the implementer landscape is pretty flat: some individual contributors are recognized as more experienced or competent than others, and their pay and task assignments often reflect that, but there's not much difference in organizational authority. In some companies, however, there are career ladders for engineering paths.

Considering Company Culture

Decide if your interest lies in coding, programming and implementation or in managing teams and outlining goals. Evaluate the company culture to ensure it fits with this preference. At XYZ Engineering, there may be an explicit non-management career ladder and the difference between the higher and lower rungs has to do with design authority: as you get more senior, you're given more decision-making power over what gets implemented and how it gets implemented. Managers are focused on business need prioritization, interpersonal relationships, and ensuring access to resources.

2.5.4 *Location Matters*

Booming locations for Data Science and analytics talent include: New York-Newark-Jersey City, Chicago-Naperville-Elgin, Washington DC-Arlington-Alexandria, LA-Long Beach-Anaheim, San Francisco-Oakland-Hayward. China is accelerating quickly as a quantitative powerhouse, with not only investments in US companies, but also a booming local tech scene with companies like Baidu, Alibaba, and Tencent [36]. With access to troves of data and the second largest economy in the world, be on the lookout for Chinese firms to snatch up talent from the US.

There are some upsides in being in a high demand market but also there is the risk of being washed away in the tide, especially when first starting out. These markets can be as over-saturated as they are large with a high volume of jobs available, but even more applicants competing for those jobs.

Malcolm Scott took a job in New York, and eventually moved to a more senior job also in New York. When it came time to leave that job it became clear one of the downsides to working in a major hub. His choices were to either (1) take an entry-level position in a different company with its entry-level salary simply because there was so much supply that he couldn't demand any more for an individual contributor position, (2) make a career change and become a full-time manager, or (3) move to a place where the market wasn't so saturated.

Implications of Location
Initial job placements may be easier in large market cities as the number of entry level jobs are higher, however, as your career progresses the competition will become intensified as fewer and fewer management and leadership roles are available. Places like the Bay Area and New York City are in the highly competitive tier. Currently, places like Chicago, Atlanta, Austin, Washington DC, Denver and the North Carolina Research Triangle could be good choices for job opportunities with intermediate management competition. Next, places like Columbus, Salt Lake City, Indianapolis, Nashville, and others are slightly smaller markets, but have tech industries that are really large compared to the rest of the country. These markets may not have as many experienced people as they need, partially because the highly competitive locations are attracting experienced people.

2.5.5 *Who's Hiring*

Currently, the highest number of openings are in three sectors: finance and insurance, information technology, and professional, scientific, and technical services [37]. The 2018 Burtchworks Data Science Salary study found tech and financial services industries to be the largest employers of data scientists (54% combined) [38], however, more opportunities are cropping up across industries including Industrial, Retail, Healthcare, and Agriculture. Also see Sect. 2.2 for in demand applications and companies doing work in those areas.

The beauty of Data Science is that there are opportunities in almost every sector, if you look for them, but some sectors are more prominent. The benefit of having niche skills is that you have an opportunity of getting in on the ground floor and making a big impact and a name for yourself.

Sectors Making Strides As health records move from paper to digital, there are some big issues to consider, not least security and data protection. Long-term, however, the potential to target healthcare services to meet demand and pinpoint health trends that might not have been spotted, are compelling. With copious amounts of messy data, new techniques, and an increasing number of data sources (including everything from patient records to x-ray images to fitness tracker data), the opportunities for data scientists with healthcare experience are substantial and growing [39]. Most notably, these opportunities are not simply limited to Silicon Valley start-ups but taking hold in more established corporations across the US. This is leading to increased geographic variability in the Data Science market with some companies even building out remote-based Data Science teams.

Industries like Agriculture or Manufacturing are also making great strides in the direction of Data Science. Monsanto shook up the Agriculture world by purchasing a big data weather company, Climate Corp, for nearly $1 Billion dollars. Companies like United Healthcare and Geico Insurance are now automating things like detection of fraud, waste and abuse in the insurance system. There are huge opportunities for disruption in traditional spaces as well, so start-ups or research teams within larger companies, such as Stanley, and Black & Decker's Digital Accelerator, can be viable options to innovate. Recently, Coursera put out a global skills index study and ranked the top 10 industry sectors, that encompass many of the largest companies globally, on their readiness for the Fourth Revolution or the Data Revolution [40]. Based on their skills landscape, the industries that are best positioned today to take advantage of this emerging environment are Manufacturing, Technology, and Telecommunications as they consistently lead in the rankings across Business, Technology, and Data Science. Industries in the middle tier were Consulting, Media, Healthcare, and Insurance who need more skilled people to stay competitive. The bottom tier were Finance, Automotive, and Consumer Goods who had a severe lack of people needed. This could explain, in part, the rush for data scientists in the finance sector. Do you want to be able to grow from peers and mentors or blaze a trail for an emerging sector in data science? For career seekers, all of the sectors above need qualified people. If you go in with your eyes open, you can get what you want!

References

1. 2018. *Data Scientist Skills*. Retrieved from https://www.datasciencedegreeprograms.net/skills/
2. 2018. *Data Science Jobs*. Retrieved from https://www.datasciencedegreeprograms.net/jobs/
3. 2018. *Job Profile: Data Engineer*. Retrieved from https://www.datasciencedegreeprograms.net/job-profiles/data-engineer/
4. Evans, C. (2018, August 27). *Beyond the unicorn: 4 developing data scientist career paths*. Retrieved from https://www.burtchworks.com/2018/08/27/beyond-the-unicorn-4-developing-data-scientist-career-paths/
5. Tyka, M. (2015, June 19). *These are what the Google artificial intelligence's dreams look like*. Retrieved from https://www.popsci.com/these-are-what-google-artificial-intelligences-dreams-look
6. Bergstrom, C. and West, J. (2017). Retrieved from https://callingbull.org/
7. Kwartler, T. (2017, February 4). *How AI is chaing the way we assess vehicle repair*. Retrieved from https://venturebeat.com/2017/02/04/how-ai-is-changing-the-way-we-assess-vehicle-repair/
8. Bresnick J (2018), *Top 5 Use Cases for Artificial Intelligence in Medical Imaging*. Retrieved from https://healthitanalytics.com/news/top-5-use-cases-for-artificial-intelligence-in-medical-imaging
9. Rayome, A. (2018, February 15). *How Sephora is leveraging AR and AI to transform reatil and help customers buy cosmetics*. Retrieved from https://www.techrepublic.com/article/how-sephora-is-leveraging-ar-and-ai-to-transform-retail-and-help-customers-buy-cosmetics/
10. Lambert, J. (2019). *Computer vision companies*. Retrieved from http://www.lengrand.fr/computer-vision-companies/
11. (2017). *Pyimage Jobs*. Retrieved from https://jobs.pyimagesearch.com/ref/3/
12. (2019). Lucidworks Customers. Retrieved from https://lucidworks.com/customers/

13. (2012). Lucidworks Compaies Using Lucene/Solr. Retrieved from https://lucidworks.com/2012/01/21/who-uses-lucenesolr/
14. (2019). Elasticsearch on stackshare. Retrieved from https://stackshare.io/elasticsearch/in-stacks
15. McClelland, C. (2017, December 4). *The Difference between artificial intelligence, machine learning, and deep learning*. Retrieved from https://medium.com/iotforall/the-difference-between-artificial-intelligence-machine-learning-and-deep-learning-3aa67bff5991
16. Ulbert, S. (2015, September 15). *The difference between predictive maintenance and preventative maintenance*. Retrieved from https://www.coresystems.net/blog/the-difference-between-predictive-maintenance-and-preventive-maintenance
17. Ulbert, S. (2015, September 15). *The difference between predictive maintenance and preventative maintenance*. Retrieved from https://www.coresystems.net/blog/the-difference-between-predictive-maintenance-and-preventive-maintenance
18. Giang, V. (2013, March 14). *Companies are putting sensors on employees to track their every move*. Retrieved from https://www.businessinsider.com/tracking-employees-with-productivity-sensors-2013-3
19. (2019). *Companies using Iot*. Retrieved from https://www.silicus.com/lp/e/azure-iot-development-services/?pi_campaign_id=63186&utm_campaign=Azure%20IoT%20Services&utm_source=PPC-Adwords&utm_medium=cpc&utm_term=Internet%20of%20things%20consulting&pi_ad_id=250023919173&utm_content=Internet-Of-Things-Consulting&gclid=Cj0KCQjwrszdBRDWARIsAEEYhrcHS-RgqLvIr79bUcotqS55sGyOaE_x3EJ9H-_ysMM0yEy_gVKpTLwaArTrEALw_wcB
20. (2018). *Data science project scoping guide*. Center for Data Science and Public Policy – The University of Chicago. Retrieved from https://dsapp.uchicago.edu/home/resources/data-science-project-scoping-guide/
21. Wheele, L. (2015, March 3). *CDC: 1 in 6 truck drivers doesn't wear a seat belt*. Retrieved from http://thehill.com/regulation/234455-1-in-6-truck-drivers-dont-wear-a-seat-belt-cdc-says
22. Suda, B. (2018). *2017 Data Science Salary Survey*. O'Reilly Media, Inc.
23. National Academies of Sciences, Engineering, and Medicine. (2018). *Data Science for Undergraduates: Opportunities and Options*. Washington, DC: The National Academies Press. https://doi.org/10.17226/25104
24. Rawlings-Goss. (2018). *Keeping Data Science Broad: Negotiating the Digital & Data Divide*. National Science Foundation. Retrieved from https://drive.google.com/file/d/14l_PGq4AxOP9fhJbKqA2necsJZ-gdiKV/view
25. Gould, R., Peck, R., Hanson, J., Horton, N., Kotz, B., Kubo, K., Malyn-Smith, J., Rudis, M., Thompson, B., Ward, M., Wong, R. (2018). *The two-year college data science summit Report*. https://www.amstat.org/asa/files/pdfs/2018TYCDS-Final-Report.pdf
26. (2019). *Data science certificate*. Retrieved from https://www.extension.harvard.edu/academics/professional-graduate-certificates/data-science-certificate
27. Business-Higher Education Forum. (2018). *Building a diverse cybersecurity talent ecosystem to address national security needs*. http://www.bhef.com/sites/default/files/2018BHEFUSMCaseStudy.pdf
28. Business-Higher Education Forum. (2018). *Building a diverse cybersecurity talent ecosystem to address national security needs*. http://www.bhef.com/sites/default/files/2018BHEFUSMCaseStudy.pdf
29. Business-Higher Education Forum. (2017). *Investing in America's data science and analytics talent*. PWC Business Higher Education Forum. http://www.bhef.com/sites/default/files/bhef_2017_investing_in_dsa.pdf
30. U.S. Census Bureau. (2017, October 24). *Employment Projects: 2016-26 Summary* (Report USDL-17-1429). https://www.bls.gov/news.release/ecopro.nr0.htm
31. U.S. Census Bureau. (2017, October 24). *Employment Projects: 2016-26 Summary* (Report USDL-17-1429). https://www.bls.gov/news.release/ecopro.nr0.htm

32. Burning Glass Technologies (January 2017). Job post estimates include actual job growth, job replacements, and churn
33. Business-Higher Education Forum. (2017). *Investing in America's data science and analytics talent.* PWC Business Higher Education Forum. http://www.bhef.com/sites/default/files/bhef_2017_investing_in_dsa.pdf
34. Columbus, L. (2017, December 11). *LinkedIn's fastest-growing jobs today are in data science and machine learning.* Retrieved from https://www.forbes.com/sites/louiscolumbus/2017/12/11/linkedins-fastest-growing-jobs-today-are-in-data-science-machine-learning/#1ae629151bd9
35. Wheeler, S. (2018, August 28). *Data science career advice to my younger self.* Retrieved from https://towardsdatascience.com/data-science-career-advice-to-my-younger-self-4c37fac65184
36. Chen, L. (2017, November 1). *Alibba, Tencent pressured to live up to $450 billion rally.* Retrieved from https://www.bloomberg.com/news/articles/2017-11-01/alibaba-tencent-face-pressure-to-live-up-to-450-billion-rally
37. Business-Higher Education Forum. (2017). *Investing in America's data science and analytics talent.* PWC Business Higher Education Forum. http://www.bhef.com/sites/default/files/bhef_2017_investing_in_dsa.pdf
38. Burtch, L. (2017). *The Burtch Works Study: Salaries of Data Scientists.* Burtch Works, Executive Recruiting. http://www.burtchworks.com/wp-content/uploads/2017/05/DS-2017-Industry.pdf
39. Agrawal, S. (2017, October 8). *Why hospitals need better data science.* Retrieved from https://hbr.org/2017/10/why-hospitals-need-better-data-science
40. Coursera (2019). Global Skills Index. Retrieved from https://www.coursera.org/gsi

Chapter 3
Building Data Programs

3.1 A GPS for Learning and Work

Data Science degrees, programs, and initiatives are emerging at a rapid pace at universities and colleges in the U.S. and abroad. Data Science, however, is as much a practice as it is a discipline, raising the questions of whether and how Data Science should be treated in academia. Should it be its own major, department, or division at a university? And what are the foundational elements that comprise a degree in Data Science? Here, we discuss the institutional barriers to developing and implementing Data Science/Analytics programs, the role of faculty, and resources for curriculum. We also feature different models taken by top U.S. institutions for incorporating Data Science on campus, how to access data, as well as potential solutions to the common challenge of faculty burden. Lastly, we go through top recommendations produced by national forums, with hundreds of researchers, convening to discuss these topics.

The modern world is in the midst of a "Data Revolution". Businesses are booming as a result of new technologies perfected for both utility and speed. This Data Revolution, however, is similar in a lot of ways to the Industrial Revolution. In the Industrial Revolution new technologies led to new jobs, brought in billions to the economy, and sparked the manufacturing dominance of certain nations and companies for the next 100 years. The same can be true of data. The conditions that positioned nations to take a dominant role in the Industrial Revolution was unpacked by the economist Joel Mokyr who made the compelling argument that the reason for the explosion of innovation during the Industrial Revolution was the number of skilled everyday workers who were comfortable using the new tools and machines of the day [1, 2]. People were trained to be comfortable with manufacturing tools, so they were able to try new things, innovate, and perfect manufacturing processes across the country. In essence, a "manufacturing literate" society equaled an innovation rich country.

R. Rawlings-Goss, *Data Science Careers, Training, and Hiring*, SpringerBriefs in Computer Science, https://doi.org/10.1007/978-3-030-22407-3_3

Today, we do not have "data literacy" society for this new data revolution. The United States and the world face a lack of skilled workers with the data literacy to explore the boundaries of the new tools being created. Tools to optimize the huge flows of data coming from companies, universities, and cities are not a part of the general educational system. Data that can be used to benefit health, energy, manufacturing, and the environment lay unused because of a lack of workforce and managers with the skills to understand the data, its tools and it uses. There are so few workers that it poses a potential danger to any nation's growth and dominance in this era [3]. The majority of the destabilizing Data Science and analytics skills require significant time and effort to develop, this is exacerbating the talent shortage. Heightened education and experience requirements are being requested for many new and emerging positions [4]. That is why many have stressed the need for increasing 'data literacy' in both the U.S. and globally. Calls for improved data literacy across industry have been happening for over a decade [5–9]. The United Nations report "A World That Counts" [10] called for "a special investment to increase global data literacy", notably to address capacity gaps [11]. Mitigating these gaps will take a concerted effort from a number of stakeholders, and there is no one-size-fits-all solution.

With the stakes this high, Data Science degree programs, must take on the mindset of acting not only as an imparter of critical skills, but as a GPS for learning and work. Students need skills, but also direction, in the sea of choices presented. Identifying the hardest-to-fill and fastest growing Data Science and analytics jobs and skills is critically important for educators, employers, and other stakeholders, thus making it a ripe area for collaboration [12].

3.2 Institutional Culture

Institutional Barriers to Developing and Implementing Data Science/ Analytics Programs

Academic institutions are slow-moving entities, for good reasons, but in fast-moving fields like Data Science academic culture can be an obstacle. The reason is that universities were not really built for disciplines like Data Science. The past 100 years of education have worked by creating smaller and smaller sub-specialized departments. For example, biology has forked into cell biology, computational biology, biochemistry, biophysics, and many others. Data science requires a reversal of this process breaking traditions and teaching a mixture of things from different fields. This can include such disjoint fields as statistics, math, computer science, ethics, business intelligence, systems engineering, graphic design, communications, and many other fields that do not typically "talk" to each other. Therefore, where Data Science should live in the university structure is up for grabs and there is no one right answer.

> Those who see it (Data Science) as a new, emerging field see Data Science as drawing on methods from many existing fields, e.g., computer science, mathematics, operations research, and statistics. Others see Data Science as simply an evolution of statistics, e.g.,

> anticipated as early as 1962 by John Tukey, or an evolution of computer science, e.g., as probabilistic and statistical reasoning becomes as important as symbolic and logical reasoning in computing. Regardless of whether Data Science is "new" or not, there was consensus that concepts and techniques from (at least) computer science and statistics are core to Data Science.—Data Science Leadership Summit Report [13]

The inherently siloed and bureaucratic structure of most universities creates a set of institutional barriers to the development of sustainable Data Science/Analytics programs. Some perceive Data Science/Analytics as a threat to "business as usual". Others try to claim sole ownership of the field and exclude other disciplines. There is often a lack of understanding by administrators about the benefits and the resources needed for Data Science/Analytics programs. Added to this are the institutional barriers created by the relationships of the departments/colleges with the central administration. Some administrations may even view Data Science/Analytics degrees as a way to increase revenue quickly, but not as a viable sustainable long-term degree program.

Additionally, slow and layered degree program approval processes discourage faculty and departments from embarking on the development of new programs. Some states can add additional layers to this process by involving other universities in the approval process. A final barrier is the difficulty in obtaining qualified faculty for Data Science programs. In times of tightening university budgets, retiring tenured faculty are not being replaced with new tenure-track faculty.

3.3 Interdisciplinary Collaborations (The Role of Faculty)

Although the jury is still deliberating on the boundaries of content areas within Data Science, there are areas that are emerging as core: statistics, modeling, programming, data mining, machine learning, visualization, ethics, research design, databases, algorithms, parallel computing, and cloud computing. Current faculty that consider themselves to be in the Data Science space tend to hold degrees in disciplines related to the aforementioned content areas. These faculties are central to developing, disseminating, facilitating knowledge acquisition, and advancing Data Science theories, principles, applications, and research.

Due to the cross-disciplinary nature of Data Science, working across the traditional disciplines would be desirable at most institutions. Regardless of the departmental structure selected to house a Data Science degree program, faculty from different disciplines will have to collaborate in ways that faculty have not historically operated. As a result, many questions will surface. Will tenure and promotion criteria change with the advent of more collaborative research outputs? How will existing faculty retool themselves through professional development, so they are adequately equipped to effectively contribute to teaching and research in the field of Data Science and the other academic disciplines that will benefit from and/or be impacted by the applications derived from Data Science, such as Health Informatics, Computational Chemistry, Digital Journalism, Computational Social Sciences, and other data-centric academic disciplines?

3.4 U.S. Models for Data-Focused Programs

The question of how Data Science fits into a university was recently discussed in at the Data Science Leadership Summit [14], where experts running Data Science programs from the top U.S. schools gathered to compare notes and discuss their schools' models for tackling Data Science. Harvard, Yale, Berkeley, University of Michigan, Georgia Tech and many others all came together to talk about what they were doing and to work through this question "How does Data Science Fit into a University?" The answer after two workshops and several days was "One Model Does Not Fit All". While "One model does not fit all" seems general it is actually a sound well thought out realization that could be the foundation for change. Realizing this gives room for flexiblity and distinct styles to emerge.

There were five distinct styles commonly taken toward Data Science within universities (and a sixth which is not commonly taken). There are pros and cons to each method for students and employers:

- Building a New Academic Unit
- Expanding an Existing Unit
- General Data Literacy for All
- Creating New Connectors
- Creating a new stand-alone entity
- Provide a Data Residency or Exchange Program

3.4.1 *Building a New Academic Unit*

Building a new academic unit such as a School of Data Science or a Department of Data Science has the benefit of being able to design a complete curriculum targeted toward Data Science training. The coursework, if done right, will be carefully selected to provide foundations in data literacy, coding, statistics, ethics, and tools. It also confers the distinct label of Data Science to students on their degree, thus, making it clear to employers the title of data scientist is appropriate.

In building any academic school or department decisions must be made that are potentially hard to change down the road, particularly for undergraduate education. For instance, a university may get locked into a particular sequence of courses with no room to add new advances or ideas. New faculty may need to be hired who will have certain hard-to-find expertise. Also, students run the risk of name creep. While the principles of analyses of data will not go out of fashion, Data Science may not be the buzzword for the topic it is today. It might be Big Data, or Information Wrangling in 10 years, but their degree will always be Data Science.

Despite the challenges, the benefits are immense. It allows companies and partners to more easily find the programs to offer sponsorship, internships and support. Case in point, the recent $120 million dollar gift to the University of Virginia, the largest in school history, came in 2019 to start a new Data Science program [15].

3.4.2 Expanding an Existing Unit

Some schools opt to expand an existing department, usually computer science, statistics or math departments, to include Data Science. This inclusion can be formal, such as officially changing the name of the department. It can also be informal where one department just takes the lead in the Data Science efforts across campus.

The benefits to this approach are longevity. Existing departments have stable budgets and footing within the university. They also have faculty and can expand more quickly than a brand-new Data Science department may be able to grow.

One consequence of Data Science success is that many departments may want to own or have pieces of Data Science live within their purview. This could be good, creating multiple places for students to thrive and grow in data acumen. It could also be detrimental, if curriculum is scattered or incomplete. Naturally, a statistics department that is now hosting Data Science would skew more toward the statistical foundations of Data Science because of the faculty expertise and traditional student body.

Another pitfall of a traditional department owning Data Science is that it runs the risk of repeating the demographics of the host department. For instance, social science students interested in Data Science may be dissuaded from participating in a Data Science program hosted in a math department, particularly if they do not see themselves as traditional math students. Disciplines with historically low numbers of women and minorities may repeat those trends for Data Science at their school. For instance, what would the difference in student body be if the Data Science program was hosted by the biology department vs. computer engineering?

This phenomenon can also be restated as an advantage. Because Data Science is so broad, it presents an opportunity to host Data Science directly within disciplines with broader student demographics as well as to expand the breadth and portfolio of existing departments. A great example of this is Yale, who changed the name of their statistics department to the Yale Department of Statistics and Data Science [16].

3.4.3 Data Literacy for All

Some universities have taken the approach that all students across all majors need some foundational data literacy for the twenty-first century workforce. This idea is borne out of the 2018 National Academy of Science report on *Envisioning the Data Science Discipline*.

> A critical task in the education of future data scientists is to instill data acumen (or data literacy).—NAS 2018 [17]

To achieve this goal some institutions have developed *Introductory Data Science* courses open to all students that can be counted as an introductory math or science elective. This opens the door to more students across disciplines getting some form of Data Science experience and training that could interest them in the topic or benefit them down the road. Because the majority of students will only be exposed to one Data Science class, it puts a lot of burden on the class to be comprehensive. Also, if done in isolation, this method provides broader exposure but without the opportunity to delve into deeper knowledge and mastery.

This is, however, an excellent tool for building up the data literacy across campus not only among students but among faculty and administrators. It can help to educate faculty across campus as to the benefits of Data Science in their fields of study, and for their student body. It can also demonstrate for administrators the high level of interest in the topic, leading to support for bigger initiatives. One example that stands out is U.C. Berkeley and their Data 8: Introductory Data Science course, which is now open to all Berkeley freshman and currently enrolling over 1200 students per semester [18].

3.4.4 Creating New Connectors

Many institutions are now seeing the potential for Data Science as a tool for modern scientific discovery. This view makes it clear that most scientific disciplines are using or can benefit from some subset of Data Science techniques to further and accelerate the discovery process. Departments are then creating Data Science connector courses that specifically look at the application of Data Science to their chosen domain. This concept is sometimes called Data Science + X, where X is any other discipline. For instance, courses such as Data Science for Art History, Data Science for Chemistry, and Analysis of Ecology Data would illustrate this point.

This idea of connectors has been extended to offering joint degrees such as Computer Science + Math, or Data Science + Ecology. For courses, this idea is not unique to Data Science. Computer science and the label computational science have been used in this manner.

One example is the University of Illinois who has a number of connector degrees between Computer Science and now potentially Data Science with disciplines like Anthropology, Linguistics, and Chemistry. One downside is the distinction or overlap between different connector type degrees is unclear, such as the difference between a Computer Science + Ecology vs. Data Science + Ecology vs. Computational Ecology degree. Each, one would assume, would have a slightly different focus but the overlap is not defined.

A distinct advantage of this model is that the Data Science methods included in a connector degree are more focused and of immediate use to the students. It makes the addition of new learning concepts also more straightforward. New Data Science lessons are added only as they become useful to new challenges in the field or help speed up the scientific discovery process. This connector model, if attached to concrete challenges, has the symbiotic effect of suggesting new tools that are needed within Data Science and advancing the cause of science in the connector area.

3.4.5 Creating New Stand-Alone Entity

Data Science Institutes/Centers/Initiatives that are not tied to any academic unit have sprung up at many of the top U.S. universities. They recognize the need for broad engagement across campus. Data Science is impacting most if not all academic departments, as well as campus life, administration, facilities, and student services. This comprehensive view of Data Science as a campus resources, as well as an interdisciplinary institute is profound. Stand-alone institutes, depending on where they sit within the university, cast a wide net and can act as a neutral convener. They can also provide a central face for Data Science activities across campus, acting as a one stop shop for faculty, administrators, students, as well as industry and government partners. The power of this model is in the ability to coordinate around common goals, keep track of national trends, and facilitate larger engagement experiences.

The downside is that by themselves institutes and centers do not provide the in-depth type of degree-based training found in the other engagement types. Regardless, they can be an asset, particularly if used in conjunction with other degree-based offerings.

An example of this is the Georgia Tech Institute for Data Engineering and Science (IDEaS) [19].

3.4.6 Data Residency or Exchange Program

Because Data Science is so varied and ever changing, it is difficult, with a limited number of credits, to prepare students for all they might encounter in the workforce. For instance, to analyze medical images requires vastly different tools and training compared to the analysis of twitter feeds in a disaster response. Detecting stock market trends require different tools than predicting genomic drug design, but students from Data Science programs could be employed in all of these fields. It then becomes difficult to have one program type that addresses all of these needs.

Therefore, I return to the metaphor of medicine from the introduction to this book. Data science is a practice as much as a discipline. Therefore, a residency program may be needed. A residency program would allow students to learn a multitude of subject areas before they specialize, which could expand the scope of data education. The infrastructure of teaching hospitals attached to medical schools allows for hands-on practice before venturing out alone. Although, there are currently some programs that are taking steps in this direction. A career seeker must decide what style of data-enabled student one wants to be in order to choose the best program.

Ideally, pre-data science programs, like pre-med programs, could prepare students for more intensive training in the rigors of the modern scientific and business workplace. A formal applied rotation would provide a well-rounded foundation in the tools and ethical considerations needed for different applications of Data Science. Admittedly, a full consensus has not been reached on what is definitively working in each field, but professional societies and domain authorities are taking on this topic. For example, the Materials and Manufacturing think tank, TMS, is taking on a Materials and Manufacturing data workforce skills study. This will include industry, but also the scientific community, to get a handle on the current sets of skills needed in the market today as well as projecting into the future for training and education relevant to sustained progress in their field.

One method of achieving this goal would be to allow for exchange programs among different academic units for Data Science students. This is also achieved in part through capstone projects with industry, hack-a-thons, or challenges. These give students exposure to different areas but do not usually provide the deeper dive in to concepts and best practices for use. Hack-a-thons leave it more to the participants own creativity to devise techniques, which has two sides. It has the downside that no form of best practice transfer is taking place, so similar mistakes and misconceptions can be learned by all. It adds the benefit, however, of being engaging and teaching independent thinking, which is critical for a data career.

3.4.7 Career Services and Support

Not every student has a relationship with a faculty member or advisor that will yield a job. Many students will go to career services to find opportunities. Career services staff who work with students in data related majors should be educated on the skills and needs of the workforce to better serve students and act as a connector to smaller organizations who may come to the career services office for assistance in finding talent.

3.5 Resources for Data Science Curriculum

Curriculum is absolutely central to any discussion of an evolving field.

The discussions surrounding Data Science curriculum are not unique to one type of institution. Higher education institutions at all levels face challenges in creating productive transitions between various academic levels and programs. While curriculum is traditionally the purview of educators, particularly teaching faculty, it is important to remember that many other groups should have a say in how it is developed. Students, as the recipients of and participants in the curriculum, have a huge stake in seeing the content and approach be effective, relevant, and dynamic. The employment sector, including small business, large industry, nonprofits and government agencies, have an equal interest in the outcomes.

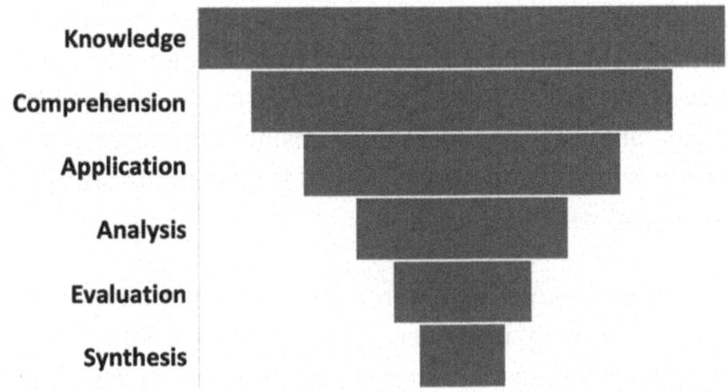

Fig. 3.1 Bloom's taxonomy as augmented by the E.U. EDISON Project Model Curriculum framework: Learning levels and action verbs [31, 32]

A few publications have begun to outline what a full-fledged curriculum might look like [27–29]. Also, the European Data Science Academy [30] publishes extensive curricular materials for a wide range of courses.

3.5.1 Data Science Module Curriculum: Learning Levels

Bloom's taxonomy provides a way to organize levels of learning, and assigns action verbs to each level for instance, remembering facts, understanding those facts, applying knowledge, analyzing, evaluating, and creating. These help to understand activities related with particular level of learning (see Fig. 3.1).

The E.U. EDISON project did a two-year study of Data Science industry needs and came up with a model for Data Science learning based on similar key levels: Knowledge gathering, Comprehension, Application of Knowledge, Analysis, Evaluation, and Synthesis. For instance, students start at the knowledge level when they can name and identify relevant technologies. They then move to the comprehension level when they can explain how technologies work. They then move to the application level when they can choose the right technology to solve a problem. Lastly, they can progress to analysis, synthesis, and finally evaluation levels.

Below are examples showing typical attributes of the different levels of learning and example questions to test these levels.

Knowledge Exhibiting memory of previously learned materials by recalling facts, terms, basic concepts and answers. *Questions like*: What are the main benefits of implementing data analytics methods for an organization or a general research group?

Comprehension Demonstrating understanding of facts and ideas by organizing, comparing, translating, interpreting, describing, and stating the main ideas. *Questions like*: Compare the business and operational models of private clouds and hybrid clouds?

Application Using new knowledge. Solving problems in new situations by applying acquired knowledge, facts, techniques and rules in a different way. *Questions like:* What data analytics methods should be applied for specific data types, analysis, or for specific business processes and activities? Which Big Data services architecture is best suited for a medium sized research organization or company, and why?

Analysis Examining and breaking information into parts by identifying motives or causes. Making inferences and finding evidence to support generalizations, and analysis of elements, relationships, organizational principles. *Questions like*: What data analytics methods and services are required to support the typical business processes of a research university? Give suggestions on how these services can be implemented with the selected data analytics platform, including on-premises or outsourced to cloud. Provide references to support your statements.

Synthesis Compiling information together in a different way by combining elements in a new pattern or proposing alternative solutions. Production of a unique communication, a plan, or proposed set of operations, derivation of a set of abstract relations. *Questions like:* Describe the main steps and tasks for implementing data analytics and data management services for an example company or teaching organization? What research and student projects can be moved to clouds and which should remain in on campus facilities and run by university personnel?

Evaluation Presenting and defending opinions by making judgments about information, validity of ideas or quality of work based on a set of criteria, either in terms of internal judgments or external evidence. *Questions like:* Do you think that implementing a data lake model for facilities across campus creates benefits for employees, short term and long term?

This framework can help progress curriculum as programs think through how to progress students through these levels. The E.U. EDISON report was also focused on developing model curriculum frameworks for different types of Data Science roles specifically Data Science & Data Analytics, Data Science & Data Management, Data Science Engineering, Data Science Research Methods, and Business Process Management. Refer to the original report [33] for details.

3.6 Continuous Instructor Learning

3.6.1 Faculty Career Advancement

Faculty are considered the conduit for knowledge dissemination, and utilization. Therefore, not only should faculty teaching in Data Science programs engage in training, faculty from all disciplines that are interested in the application of Data Science to their fields and professions should participate in Data Science training as well. The development of interdisciplinary application-based Data Science training would also advance the career of a tenure track faculty member by increasing the number a quality of publications, opening up new avenues of explorations as well as new sources of funding, and facilitating interdisciplinary collaborations.

3.6.2 Benefits of Interdisciplinary Collaborations

All faculty can benefit from Data Science by being able to more effectively train their students to be workforce ready as well as being able to make more meaningful research contributions that are informed by both Data Science and domain specific phenomena including, but not limited to, astronomy, biology, business, chemistry, environmental science, medical data, political science, physics, social sciences, behavioral sciences, the arts and humanities. For this reason, an introduction to discipline-specific Data Science tools should be embedded in all disciplines. For example, multi-level courses could be cross-listed, and team taught—an art course that teaches "Image Processing" could be an advanced art class, but also a computer science elective/topics course. Such embedded courses would truly foster interdisciplinary collaboration.

Institutions will have to decide the appropriate departmental structure for their Data Science or related degree programs, the tenure and promotion criteria for faculty teaching in the Data Science program, faculty credentialing criteria to meet accreditation standards, how to provide professional development to support the contributions that faculty can make to the field of Data Science, how to provide training in the areas of pedagogical best practices and team teaching, and how to partner with industry and other stakeholders for curriculum, student, and faculty development.

3.6.3 Faculty Training and Credentialing

To address the challenge of faculty who do not have the expertise and need to retool, many faculty bolster their expertise and credentialing in the Data Science arena by attending industry- and academic-driven training sessions or conferences. Industry-driven training videos and classes are offered by companies such as Microsoft, IBM, NVIDIA, Intel, and PyData. Academic-driven Data Science training are available as well such as the University of California - Berkeley, New York University, and University of Washington who are offering training sessions and providing the educational materials as a result of funding from the Moore and Sloan Foundations. The National Science Foundation is also providing funding to support academic training in the area of Data Science. In an effort to advance Data Science as an academic discipline, faculty can take advantage of the existing Data Science training activities, and trainings should be made accessible online from all types of academic institutions and disciplines.

Another way to train and credential existing faculty could be through non-traditional fellowships for existing faculty. Short-term fellowship projects with industry could be structured around developing best practices. Research projects in Data Science pedagogy as well as the application of Data Science to the faculty member's discipline are good tools. This provides expertise in dealing with data throughout the data life cycle and within the Data Science ecosystem.

3.6.4 Faculty Recruitment

Faculty recruitment and retention in a purely Data Science school is challenging due to competition from industry, and other top tier institutions. As talent is scarce, the university salary alone will usually not be compelling enough at even top institutions to draw all the talent necessary to meet the growing demand of students interested in this area. Universities, particularly smaller institutions, must be creative and focus on additional benefits.

An initial recommendation is to "look within" and assess the talent of your current faculty and try to augment those skills as described above. Provide clear benefits and credit toward advancement for those faculty that choose to reskill. Second, to recruit externally, highlight opportunities for national experts that join your faculty to act as trainers and host training sessions for others on campus. Third, be open to a shared faculty model of cross-departmental or cross-institutional faculty with joint appointments. One caution is to clearly define the career path of these individuals so that they do not end up beholden to the full expectations of multiple departments to succeed.

With the recent advent of master's and bachelor's level degrees in Data Science, there is a need for more doctoral level programs in Data Science and analytics. Because the spokes of the Data Science wheel are so varied, yet interwoven, faculty with degrees in related disciplines previously delineated would continue to be more prevalent until universities start offering doctoral degrees in Data Science. Therefore, over the course of the next decade or so, there may be a paradigm shift in faculty credentialing for the Data Science academic field.

3.6.5 Collaborations with Industry and Government

Collaboration between academia and industry in Data Science will be critical, particularly during this phase of rapid growth and change.

There are a number of different paths to foster collaboration with industry, including experiential learning/capstone projects (e.g., solving real world problems for real companies) that improves the learning process, inviting industry employees to give guest lectures to open the door to interactions with students and faculty, and considering faculty exchanges with company experts, as well as philanthropic gifts to the academic units for data-focused scholarships. Also, involving industry in creating workforce talent pipelines and listening to their needs and desires will involve them in content development.

It is important for academics to be proactive. A lack of opportunities for students at non-research-intensive institutions to engage with industry and workforce systems can create a data divide. Academia tends to promise industry the best and brightest students as the outcome of collaborative efforts. From an industry perspective the expectation is that academic institutions should create a business case. For smaller schools especially, it is important for an academic program to be able to build a business case for long-term investment in the program. With industry, unlike government, the general need is to start small and build.

Finally, collaboration between academic programs and government workforce systems (primarily serving the underserved) is key. While the relationship between specific companies and academic institutions does create workforce pipelines, it does not produce all of the data professionals needed for the workforce. It is important for academia to gain insights from local and national government workforce boards about the skills and experiences that are needed in the workforce. The goal is two-fold to educate students and to prepare them to go into career pathways.

3.7 Access to Data

One issue that arises is access to high quality data and examples for teaching. The ability to locate useful and useable data and the credentials needed to acquire, assess, and glean knowledge from that data is lacking for classroom instruction. Given that we are living in an era with data as a part of the social and economic fabric of daily life, both aspects of "data access" are critical in education programs.

Data Wrangling 101: A Virtuous Cycle

A short course on effective 'data wrangling' would serve many students well. Professional data scientists report spending up to 80% of their time managing or wrangling data before doing any analysis and getting conclusions. Consequently, it is unrealistic for every professor to taken on the challenge of data wrangling in their individual courses or take on the burden of data wrangling for each course module.

As part of a Data Wrangling 101 course, students could take on processing, merging, and cleaning data for specific teaching purposes. Professors across campus would propose particularly relevant data topics or ask for examples of datasets to be created that illustrate different types of issues in data processing. These examples could then be used in other courses across campus, thus relieving a burden on faculty while at the same time giving students valuable experiences and a sense of ownership over what is taught at the university. Data Wrangling 202 could take requests for research groups or local industry.

Access to data is necessary for all disciplines, however, it is problematic in some disciplines like healthcare where claims data is not readily available at the patient level for students and faculty. Other datasets are expensive and contain contractual provisions that impact who may see and use the data, and how it is stored. In some cases, researchers are unable to publish their data due to legal, compliance, cost, or privacy concerns. Industry is under increasing pressure from regulators and the public to avoid hacking breaches that disclose PII (Personally Identifiable Information) or PHI (Personal Health Care Information). Thus, some critical barriers are:

- Datasets require a lot of pre-processing
- A lack of interoperable datasets
- Locating available datasets
- Actual retrieving processes
- Lack of access to local or culturally relevant data
- Issues of ethical or legal concern in accessing and using datasets
- Data quality

The volume, velocity, variety, veracity (truthfulness), and value of data, or so called 5 V's of "Big Data", can cause other issues. Some real-world datasets are so large, quickly flowing or disparate that they require a lot of time just to get them into a usable form. Developing the skills to handle the most interesting forms of data are often too time-consuming for regular classroom instruction. These datasets are closer to real life situations but can cause issues, complicating curriculum design, which can lead to instructors spending large amounts of time cleaning data or advising students on how to do so themselves. For example, when a homework or project assignment is given, in order to make the assignment relevant, an instructor may want to provide realistic data, however, finding appropriate data from multiple datasets, merging and cleaning the data can be a significant time constraint. Some resources for quick learning are cheat sheets for data wrangling in common software like R [20] and Python [21]. For already over-burdened instructors, this leads to a reluctance to include real data in coursework or to overly simplification of examples. Therefore, the question of data access has a direct translation in to the education process.

3.7.1 Trusting Data

Understanding the provenance of datasets, or how the data was gathered, is also critical for curriculum designers and teachers using readily available data. This cannot be left as an afterthought, or as part of a quickly assigned project; documentation of the origin and processing of data must include details such as:

- How the data was collected or generated, including assumptions;
- Processing methods, including methods for missing data points;
- Intellectual Property information, including ownership, licensing, and any restrictions on use;
- A persistent identifier;

Lack of these kinds of information can have a big impact on which datasets are used.

For instance, the issue of global market inflation, the United States and China measure inflation in different terms that are not readily comparable. The World Bank, International Monetary Fund, and United Nations specify how individual nations should report data to them, and it's necessary to read and understand the methodology. This is a small example of a few challenges faced with datasets.

3.7.2 Cleaning Data

Cleaning data is a particular challenge. The process must be designed to fairly reflect the underlying purpose of the analysis. Sometimes the algorithms find features of the data that are really just features of the preprocessing. Training, care, and experience are required.

The single most important best practice for cleaning datasets is that the process must be reproducible. Common steps will aid students such as 1) make a read-only archive of the original raw dataset. 2) Clean the data using scripts or programs written for the purpose, and make sure the scripts are well documented. 3) Write up the methodology used so it's easy for someone to read the code. Asking questions will influence how the data is cleaned as well as the downstream steps in a modeling pipeline [23].

3.7.3 Resources for Data

The resources listed here are examples only and their availability obscures a deeper problem.

In the last decade, state and federal datasets were made available in larger quantities. Open data portals are being created at many levels of government. DataUSA.io is the most comprehensive visualization of U.S. public data built to date and provides an open, easy-to-use platform that turns data into knowledge. State and local governments are producing their own data portals individually or through companies like Socrata who publish The Open Data Network an interface to search for similar data across multiple states and regions [22]. For example, Data.gov and the Federal Reserve Bank of St. Louis have operated the FRED data portal for at least 20 years [24]. The ALFRED data portal offers another unique look at data collected by the government, which is often useful but not well understood, even in industry. Twitter offers a developer portal [25], which allows use of the social media platform's data, and is a useful way to enrich other data sets. The World Bank, International Monetary Fund, and the United Nations are also world class sources of data.

Specific scientific areas, such as bioinformatics and genomics, are flourishing in their use of shared data. These topics are particularly interesting as a model because, for privacy reasons, not all of the data is open. The community, however, has found ways to share, helping drive down the cost and time to new discoveries. The Centers for Medicare and Medicaid Services [26] is one of the few publicly available sources of health care data.

Users of open or pre-collected data should be aware that historical data may be adjusted, corrected, or filled in. Data collection pipelines are often altered, but whether users notice or need to re-do their analysis is rarely raised as an issue. These adjustments may happen years into the future and it is up to the user to be aware of this concern. Another key problem with data collected for other purposes are adjustments to the data, whether they are seasonally adjusted, or standard adjustments are applied such as the hedonic adjustments now applied to the Consumer Price Index. Data collection methodologies could be changed over time as well. Due to changes in the U.S. economy, inflation data today, may well be different from the non-urban inflation data previously collected in the United States, and currently collected in India and China. Cross-comparisons can be made more challenging due to these types of issues.

3.7.4 Data Program Solutions

Cleaning data is essential to building high-quality models that generalize well. This task is tedious, error-prone, and not always the type of intellectual challenge that appeals to faculty. Emerging ideas, such as using machine learning techniques to clean data, are interesting and could be useful projects for undergraduates and graduate student theses and dissertations. Graduate and undergraduate students have different needs for data. In the latter case, it is helpful for students to reproduce the results of others, as closely as they can, to enhance learning. For the former, unique datasets are more important since many publicly-available datasets have already been thoroughly mined for insights. They also serve to illustrate ethical, legal, privacy, and compliance concerns as well, which furthers the goal of integrating Data Science ethics into ongoing classroom discussions. A professional teaching proper data wrangling skills and weighing in on research projects could speed the progress of not only teaching but progressing scientific knowledge as well.

Data Scientists for Universities: Relieving Faculty Burden
As an emerging best practice, academic units should consider having access to an experienced data scientist who is available to all interested faculty. They could report to a Dean or department head and assist in teaching (see Data Wrangling 101). They could also serve as an internal consultant to assist faculty with not just teaching, but research projects as well. The Goizueta School of Business at Emory University is moving toward this model. This vision allows for some centralization of data approaches, data storage, and facilitates as well as cleaning data for teaching, which may otherwise be a poor use of faculty or student time.

The Data Scientist's salary could be funded through a mix of teaching credits such as Data Wrangling 101 and 202, as well as soft funded through time put on grant projects with researchers.

3.8 Top Recommendations

For Data Science undergraduate degrees as well as two-year associate degrees, specialized governmental forums have been convened to drive toward consensus as to the elements needed to construct a good Data Science degree program. Specifically, the National Academies of Sciences, Engineering, and Medicine established the Committee on *Envisioning the Data Science Discipline: The Undergraduate Perspective*, which was tasked with setting forth a vision for the emerging discipline of Data Science at the undergraduate level. A Final Report "Realizing the Potential of Data Science" was issued from the National Science Foundation Computer and Information Science and Engineering Advisory Committee, Data Science Working Group. Also, the final workshop report of the American Statistical Association (ASA) "Two-Year College Data Science Summit". The collective recommendations from these groups are summarized below.

Undergraduate Data Science Programs: Top Ten Recommendations [34]
For over a year the National Academy of Science committee collected information from numerous stakeholders, including the author, and determined a set of ten recommendations for envisioning the data science discipline.

Recommendation 1—Faculty Development: Academic institutions should embrace Data Science as a vital new field that requires specifically tailored instruction, delivered through Data Science major and minor academic programs and requiring the development of a cadre of faculty equipped to teach in this new field.

Recommendation 2—Varied Pathways: Universities should provide and evolve a range of educational pathways to prepare students for the various Data Science roles.

Recommendation 3—Data Literacy: To prepare their graduates for this new data-driven era, academic institutions should commit to developing a basic understanding of Data Science in all undergraduates. This commitment should cover both associate- and bachelor-level degrees as well as programs in which there is no degree.

Recommendation 4—Data Ethics: Ethics is a topic that, given the nature of Data Science, students should learn and practice throughout their education. Universities should ensure that ethical considerations are woven into the Data Science curriculum from the beginning and throughout.

Recommendation 5—Ethic Code: The Data Science community should adopt a code of ethics; such a code should be affirmed by members of professional societies and included in professional development programs and curricula. The code should be re-evaluated often in light of new developments.

Recommendation 6—Institutional Bridges: Four-year colleges and two-year colleges should establish a forum for dialog on all aspects of Data Science education, training, and workforce development.

Recommendation 7—Program Diversity: As programs develop, they should focus on attracting diverse students, with varied backgrounds and degrees of preparation, and preparing them for success in a broad variety of Data Science careers.

Recommendation 8—Constant Evolution: Because these are early days for undergraduate Data Science education, academic institutions should be prepared to evolve programs over time. They should create and maintain the flexibility and incentives to facilitate the movement of courses, material, and faculty among departments and programs.

Recommendation 9—Continuous Education: During the development of Data Science programs, institutions should provide support, so the faculty can become more cognizant of the varied aspects of Data Science through discussion, co-teaching, sharing of materials, short courses, and other forms of training.

Recommendation 10—Constructive Evaluation: Academic institutions should ensure that programs are continuously evaluated and should work together to develop professional approaches to evaluation. This should include developing and sharing measurement and evaluation frameworks, data sets, and a culture of evolution guided by high-quality evaluation. One possible vehicle could be the establishment of a professional society.

Data science as a multifaceted, interdisciplinary, field of study that "focuses on the processes and systems that enable the extraction or insights from data in various forms, structured and unstructured" [35]. It touches every discipline and industry.

Two-Year College Data Science Programs: Top Seven Recommendations [36] Two-year colleges have a robust future in Data Science education. The American Statistical Association (ASA), with funding from the National Science Foundation, hosted the Two-Year College Data Science Summit. The summit assembled 72 educators, researchers and practitioners in statistics, mathematics, computer science, and Data Science. Summit participants included faculty from two-year colleges, four-year colleges and representatives from industry, government, and nonprofits.

The primary goal of the summit was to produce curricular guidelines to assist two-year colleges in establishing and maintaining Data Science programs. The participants considered three types of potential Data Science programs: (1) Associate degree programs for students who intend to transfer to a four-year institution, (2) Associate degree programs for students aiming to go directly into the workforce, and (3) credit bearing certificate programs. These three different types of programs share many aspects but differ in the emphasis placed on each of these program outcomes. Based on input from 2 days of discussions among participants, and after comparing similar curricular guidelines and suggestions from the Park City Math Institute (PCMI) Data Science Initiative and the National Academies of Sciences (NAS) Committee on Envisioning the Data Science Discipline, the conference writing team produced the following set of recommendations.

Recommendation 1—Introduce Statistics: Create courses that provide students with a modern and compelling introduction to statistics that, in addition to traditional topics in inferential statistics, includes exploratory data analysis, the use of simulations, randomization-based inference, and an introduction to confounding and causal inference.

Recommendation 2—Use Real Data: Ensure that students have ample opportunities to engage with realistic problems using real data, so they see statistics as an important investigative process useful for problem solving and decision-making.

Recommendation 3—Reduce Mathematical Barriers: Explore ways of reducing mathematics as a barrier to studying Data Science while addressing the needs of the target student populations and ensuring appropriate mathematical foundations. Consider a "math for Data Science" sequence which emphasizes applications and modeling.

Recommendation 4—Give Realistic Problems: Design courses so that students solve problems that require both algorithmic and statistical thinking. This includes frequent exposure to realistic problems that require engaging in the entire statistical investigative process and are based on real data.

Recommendation 5—Expose Students to Tools: All programs should (a) expose students to technology tools for reproducibility, collaboration, database query, data acquisition, data curation, and data storage; (b) require students to develop fluency in at least one programming language used in Data Science and encourage learning a second language.

Recommendation 6—Infuse Data Ethics: Ethical issues and approaches should be infused throughout the curriculum in any program of Data Science.

Recommendation 7—Foster Active Learning: Whenever possible, classroom pedagogy should foster active learning and use real data in realistic contexts and for realistic purposes. Programs should consider portfolios as summative and formative assessment tools that both improve and evaluate student learning.

References

1. Mokyr, J. (1985). *The economics of the industrial revolution*. Government Institutes
2. Kelly, Morgan and Mokyr, Joel and O'Grada, Cormac, Roots of the Industrial Revolution (November 26, 2015). UCD Centre for Economic Research Working Paper Series (2015), WP15/24. Available at SSRN: https://ssrn.com/abstract=2695719 or https://doi.org/10.2139/ssrn.2695719
3. Networking and Information Technology Researcher and Development Program & Data Senior Steering Group, Big. (2016). The Federal Big Data Research and Development Strategic Plan.
4. IBM, BurningGlass Technologies, Business Higher Education Forum (2017), The Quant Crunch, BurningGlass Technologies, https://www.ibm.com/downloads/cas/3RL3VXGA
5. UNESCO (2013). Literacy and competencies required to participate in knowledge societies. Conceptual Relationship of Information Literacy and Media Literacy in Knowledge Societies, 3. Research Paper from Worlds Summit on the Information Society, 2015. Paris: United Nations Educational, Scientific and Cultural Organization.

6. National Research Council (2006) Learning to Think Spatially: GIS as a Support System in the K-12 Curriculum. Report from the Committee on Geography; Board on Earth Sciences and Resources; Division on Earth and Life Studies. Washington, DC: National Academies Press.
7. Kastens, K. & Krumhansl. R. (2013). EarthCube Education End-User Workshop. Scripps Institution of Oceanography, La Jolla, California March 4–5, 2013. Arlington, VA: National Science Foundation.
8. Zalles, D. (2014). Young youth explore geospatial data for citizenship project: A case study. Menlo Park, CA: SRI International.
9. Ridsdale, C., Rothwell, J., Smit, M., Ali-Hassan, H., Bliemel, M., Irvine, D., Kelley, D, Matwin, S., & Wuetherick, B. (2015). Strategies and best practices for data literacy education: Knowledge synthesis report.
10. Independent Expert Advisory Group (IEAG). (2014). *A world that counts: mobilising the data revolution for sustainable development*. http://www.undatarevolution.org/wp-content/uploads/2014/11/A-World-That-Counts.pdf.
11. The United Nations. (2016). *Report of the eighth meeting of the statistical conference of the americas of the economic commission for latin america and the caribbean*. https://repositorio.cepal.org/bitstream/handle/11362/40065/S1600207_en.pdf?sequence=1&isAllowed=y
12. Business-Higher Education Forum. (2017). *Investing in America's data science and analytics talent*. PWC Business Higher Education Forum. http://www.bhef.com/sites/default/files/bhef_2017_investing_in_dsa.pdf
13. Wing, J., Janeja, V., Kloefkorn, T., Erickson, L. (2018) *Data science leadership summit summary report*. https://datascience.columbia.edu/files/seasdepts/idse/Data_Science_Leadership_Summit_Summary_Report.pdf
14. Wing, J., Janeja, V., Kloefkorn, T., Erickson, L. (2018) *Data science leadership summit summary report*. https://datascience.columbia.edu/files/seasdepts/idse/Data_Science_Leadership_Summit_Summary_Report.pdf
15. Anderson, N. (2017, January 18). *Record $120 million gift to U-Va. Going to hot subject in academia: Data science*. Retrieved from https://www.washingtonpost.com/local/education/record-120-million-gift-to-u-va-going-to-hot-subject-in-academia-data-science/2019/01/17/4e8e3e1c-19b0-11e9-9ebf-c5fed1b7a081_story.html?noredirect=on&utm_term=.e8f987d61b13
16. (2018). Department of Statistics and Data Science at Yale University. Retrieved from https://statistics.yale.edu/
17. National Academies of Sciences, Engineering, and Medicine. (2018). *Data Science for Undergraduates: Opportunities and Options*. Washington, DC: The National Academies Press. https://doi.org/10.17226/25104
18. U.C. Berkley (2019) *Data 8: Foundations of Data Science* Retrieved from https://data.berkeley.edu/education/courses/data-8
19. (2018). *Institute for Data Engineering and Science*. Georgia Tech. Retrieved from http://ideas.gatech.edu/
20. (2015). *Data Wrangling with dplyr and tidyr*. R Studio. Retrieved from https://www.rstudio.com/wp-content/uploads/2015/02/data-wrangling-cheatsheet.pdf
21. Lustig, I. (2015). *Data Wrangling with pandas cheat sheet*. Retrieved from https://pandas.pydata.org/Pandas_Cheat_Sheet.pdf
22. Socrata (2019). The Open Data Network Retrieved from https://www.opendatanetwork.com/
23. Kolda, T. (2017). *Sparse versus scarce*. Retrieved from www.kolda.net/post/sparse-versus-scarce/
24. Federal Reserve Bank of St. Louis. (2018). Fred Economic Data. Available from https://fred.stlouisfed.org/
25. Twitter Developer Portal. (2018) Twitter. Retrieved from https://developer.twitter.com/
26. Centers for Medicare and Medicaid Services. (2010). *Research, Statistics, Data & Systems*. Retrieved from https://www.cms.gov/Research-Statistics-Data-and-Systems/Research-Statistics-Data-and-Systems.html

27. Anderson, P., Bowring, J., McCauley, R., Pothering, G., & Starr, C. (2014, March). An undergraduate degree in data science: curriculum and a decade of implementation experience. In Proceedings of the 45th ACM technical symposium on Computer science education (pp. 145-150). ACM.
28. Cassel, B., & Topi, H. (2015, October). Strengthening data science education through collaboration. In Workshop on Data Science Education Workshop Report (Vol. 7, pp. 27-2016).
29. De Veaux, R. D., Agarwal, M., Averett, M., Baumer, B. S., Bray, A., Bressoud, T. C. & Kim, A. Y. (2017). Curriculum guidelines for undergraduate programs in data science. Annual Review of Statistics and Its Application, 4, 15-30.
30. Data Science Training and Data Science Education - EU. Retrieved January 08, 2018, from http://edsa-project.eu/
31. Demchenko, Y., Belloum, A., Wikroski, T., Cayirci, E., Krolak, A., Brocks, H., Becker, J., Manieri, A. (2015, September 1). *Edison: Education for data intensive science to open new science frontiers*. Retrieved from http://edison-project.eu/sites/edison-project.eu/files/filefield_paths/edison_d3.1_model_curricula_definition_and_report_on_the_use_cases_support_deside.pdf
32. Bloom, B. S.; Engelhart, M. D.; Furst, E. J.; Hill, W. H.; Krathwohl, D. R. (1956). Taxonomy of educational objectives: The classification of educational goals. Handbook I: Cognitive domain. New York: David McKay Company.
33. Demchenko, Y., Belloum, A., Wikroski, T., Cayirci, E., Krolak, A., Brocks, H., Becker, J., Manieri, A. (2015, September 1). *Edison: Education for data intensive science to open new science frontiers*. Retrieved from http://edison-project.eu/sites/edison-project.eu/files/filefield_paths/edison_d3.1_model_curricula_definition_and_report_on_the_use_cases_support_deside.pdf
34. National Academies of Sciences, Engineering, and Medicine. (2018). *Data Science for Undergraduates: Opportunities and Options*. Washington, DC: The National Academies Press. https://doi.org/10.17226/25104
35. Berman, F., Rutenbar, R., Hailpern, B., Christensen, H., Davidson, S., Estrin, D., Franklin, M., Martonosi, M., Raghavan, P., Stodden, V., Szalay, A. (2018). Realizing the Potential of Data Science. *ACM. 61*(4), p. 67-72. Doi https://doi.org/10.1145/3188721
36. Gould, R., Peck, R., Hanson, J., Horton, N., Kotz, B., Kubo, K., Malyn-Smith, J., Rudis, M., Thompson, B., Ward, M., Wong, R. (2018). The two-year college data science summit Report. https://www.amstat.org/asa/files/pdfs/2018TYCDS-Final-Report.pdf

Chapter 4
Building Data Talent and Workforce

Hiring data talent is desirable and challenging. Due to a global talent shortage, there are hundreds of thousands of Data Science jobs that go unfilled each year. There are, however, a few traits that differentiate the best hiring managers from the rest. Managers that take full advantage of their resources and build a culture of recruiting tend to consistently get the top candidates. In this section, we talk about why hiring for Data Science and analytics talent is different, the flaws in the traditional ways Human Resources departments are used, the strategies for building an updated culture of recruiting and the direct benefits to hiring managers. We also talk about reasonable expectations and the missing links in the hiring process, how to assess skills, talent sourcing and continuing education for current employees. Finally, we discuss considerations for senior-level data scientists and the new Data C-suite.

4.1 Why Is Hiring for Data Science and Analytics Different?

> There is a big difference between someone you would love to hire, and someone you will settle for… A lot of companies end up settling for people that are less than optimum with respect to skillset, personality, and communication skills. The major problem is not with the candidates but a real disconnect in the hiring process.—Guy Gomis, Data Science Executive Recruiter and Partner at Brainworks

Effective managers realize that the failure to hire the best talent is not solely the responsibility of Human Resources (HR) departments, recruiters, or other hiring groups to bring in top prospects. There is often a disconnect between the desired outcome and process implemented. First, competition for data talent is not solely among industry peers. Let me emphasize that, no matter what industry you are in, whether selling shoes or building planes you are going to compete with brands not only like Facebook, Google, and Amazon, but also Goldman Sachs, Bloomberg, and J.P. Morgan. Companies that are well-known for technology (like Apple or

R. Rawlings-Goss, *Data Science Careers, Training, and Hiring*, SpringerBriefs in Computer Science, https://doi.org/10.1007/978-3-030-22407-3_4

Microsoft) and those industries that typically command higher salaries (like finance) are dominating the tech talent market. The technology and financial industries combined are the largest employers of data scientists (54% combined) [1]. Second, the overall talent pool is relatively small compared to the demand. There are now thousands of top tier companies in every industry looking for data talent to improve efficiency and compete in the global market.

The education system, however, has not fully formed a pipeline to produce talent at scale without industry support (See Chap. 3). Third, the field is interdisciplinary and in continuous flux, requiring updates to current and future employee skills. Therefore, the most in demand tools of today may not stand the test of time. Finally, tasks such as producing relevant job descriptions, advertising and recruiting in appropriate places, supporting non-traditional pathways, vetting candidates, team dynamics and skills development have not been streamlined for hiring managers looking for Data Science talent.

Due to these factors and others, many data hiring managers and executives (CDO's, CAO's) feel they have been left essentially unsupported in hiring talent. A consequence of this is that data executives have been trying to develop hiring processes alongside or outside of traditional HR departments, as the traditional processes may not be producing the desired results.

4.2 Don't Go It Alone: Build a Culture of Recruiting

> A business can enjoy every advantage but not excel at hiring great analytics teams, including upper and middle managers, and chances are that business won't reach the heights it could with top talent—it may even fail due to a lack of it. Brainworks Recruiting [2]

Another shift that the most successful hiring managers are taking is to build a culture of recruiting. In such a culture, it is the responsibility of every person, manager, and executive to hire and recruit talent for the business. The shift is toward making the technical teams and recruitment professionals "talent acquisition partners". HR or recruiting firms play an invaluable role, when they become a more integrated team member through partnering with the hiring manager and the technical team. This is in stark contrast to the traditional role of recruiting in a company which has worked well for years but has some flaws for data talent.

4.2.1 The Traditional Role of Human Resources

To highlight one of the challenges in creating a culture of recruiting, let us first talk about the current best practices articulated in the hiring process from the AIRS-ADP Professional Recruiter Certification 2.0 [3]

Traditional Hiring Process

1. Initial session with hiring manager and recruiter to review profile
2. Finalize profile and sign off. Hiring manager and recruiter approve.
3. Source candidates (Recruiter's responsibility)
4. Screen candidates (Recruiter's responsibility)
5. Recruiter presents top five candidates (Recruiter's recommendation).
6. Interviews are scheduled (Recruiter arranges)
7. Interviews conducted. Recruiter, hiring manager and other key staff
8. Additional assessments conducted. Specify additional assessments
9. Final selection meeting. Recruiter and hiring manager convene
10. Close top candidate. Hiring manager closes with recruiter's assistance
11. Revise as necessary
12. Organize for start date.

Recruiter Responsibilities The recruiter is responsible for managing the hiring process. They take the lead role in finalizing the profile, sourcing candidates, organizing the assessment of candidates, orchestrating the final selection with the hiring manager, assisting in closing the candidate and conducting negotiations if required. The recruiter will then also put in place the early components of the on-boarding process. The one part of the hiring process that the recruiter totally owns and is responsible for is finding or sourcing of candidates. The other activities are done in conjunction with the hiring manager and others involved in the hiring process. The recruiter is responsible for project management, guidance, quality control and facilitation of these activities.

Hiring Manager's Responsibility The hiring manager is responsible for making the best possible hire in a timely manner. While the recruiter and others will offer recommendations on which candidate is the best fit for the job and organization, the decision is made by the hiring manager. The hiring manager also has the responsibility to participate in key recruiting activities throughout the process and to keep scheduling commitments outside of extremely urgent situations. These interim dates and commitments are important because the hiring process can become unwieldy if dates start to slip.

4.2.2 An Updated Culture of Recruiting

The problem with the traditional path is that it is a path based on exclusion (assuming there is a large pool of applicants) instead of inclusion. It does not incentivize the entire team to embrace the role of "talent scouts", constantly looking for talent to add to the team. An updated culture of recruiting creates true opportunities and benefits to managers.

Direct Benefits to Managers

1. Better Hires
2. Less Frustration with the process of hiring
3. Less Turnover on your teams
4. Less Administration of poor team dynamics or motivation of poor culture fit

The entire process of hiring great, highly sought-after, candidates requires practical steps. A partnership between the hiring manager, the recruiter, and the hiring team. The recruiter will need to be included in meetings leading up to a hire. This facilitates them not only have general knowledge about the role to be filled, but it allows them to see how the team operates as a unit.

Also, managers should not abdicate the entire responsibility of sourcing candidates to the recruiter. If a technical manager is overwhelmed by the recruitment process, I would advocate partnering on equal footing with HR in a collaborative step-by-step way to get results. Sitting down with their team and the recruiter to brainstorm where to look for talent. Then the team could come back with a shared list of candidates based on everyone's sourcing. Recruiters and hiring managers then have the shared responsibility of evaluating talent by recognizing other's strengths and weaknesses.

4.2.2.1 Talent Acquisition Partnership Guidelines for Managers

1. *Recognize those with "People Talent"*. Human Resources professionals have years of experience in reading people and assessing applicants. They can easily recognize the difference between B− talent and A+ talent on a number of levels beyond just technical prowess. They have seen hundreds and sometimes thousands of candidates over their career and the team dynamics produced. Engaging them even in the job description phase is helpful, for their insights into the type of person who would fit best in your team. Tell them about the Innovation skills (See Sect. 2.3.1) you need for your team to grow.
2. *Dig together to unearth new talent*. Again, try not to turn over the full responsibility for talent sourcing. Let recruiters show you where they will be sourcing talent and what are the best practices. Talk to your technical team to suggest new sources for candidates and look for opportunities to talk to candidates yourself.
3. *Ask questions and teach*. Inquire about the process used by HR and encourage them to ask about your work. Teach in a careful way to make them a subject matter expert on your team and talk about your current team dynamic. Team dynamic is a huge selling point, or deterrent, to Data Science applicants and something candidates ask about it directly, so recruiters should have some specific knowledge of the team. This also helps HR become a partner in creating a job description.
4. *Ask for advice*: Once you have evaluated the technical competency of top candidates ask your talent acquisition partner to sit in on interviews to weigh in on how the candidate stack up personality wise against others or any red flags they

sense during the discussion. Sometimes waiting a little longer to fill a role with the right person can make all the difference.

5. *Respect time and candidate experience*: For high-demand talent, the candidate experience will be key. Long delays, missed deadlines, and unnecessary hoops can repel top candidates from joining even an interesting team. Managers should try in earnest to keep deadlines for responding to recruits and candidates, as well as making key decisions. Waiting a long time to respond to a desirable candidate can send the message that they are not valued and should accept another offer.
6. *Work together to uncover macro-market trends*: Get feedback on what recruiters and team members are seeing market wide or in talking to candidates every day. Are there certain misconceptions about your brand? Are candidates having similar concerns about location, autonomy in the company, or career advancement? These are invaluable insights that can help leadership be aware of trends and opportunities to re-position.

4.2.3 The Foundation of Partnership

Businesses with imagination know team loyalty is the name of the game. A strong talent acquisition partnership is critical, as we have discussed, but how do you build the foundation of this partnership? It is built on 3 basic pillars:

- Mutual Goals—Make sure the team's goals are aligned to find the absolute best fit for a role. Do not take for granted that these goals will be obvious.
- Trust—Care about the success of each team member, as it promotes the success of the team. This means asking to hear from those usually quiet.
- Constant flow of communication—Make a proactive effort to remain in constant communication throughout a hiring process. This smooths any confusion that may arise.

Mainly, to gain the loyalty of your team you have to offer your loyalty as well. This means showing up with interest in the process of hiring candidates and promoting the career success of your current team. It may seem to be an extra task on your already full plate to engage in candidate sourcing. Yet, it saves vast amounts of time.

In reality, the most time-consuming result is hiring a bad team member. A bad team member can sap moral, cause untold number of management meetings to work through issues, cause the loss of existing high functioning personnel, add to documentation burden, necessitate additional review, increase the need to micromanage and to motivate constantly. This can drain the productivity of a whole team, division or company and is a real barrier to innovation. Alternatively, promoting the interests of your current team makes them ambassadors of your brand, and there is nothing stronger than a spontaneous personal recommendation from a current employee.

4.2.4 *The Missing Link*

One of the biggest mistakes hiring managers make is not spending enough time telling candidates why the position is a good fit for them. A lot of managers are still stuck in the old paradigm, in which interviewing is a one-way street. Candidates need to sell themselves on why a company should hire them. Yet many companies do not spend enough time telling candidates why they should come and work for them, not only in terms of their own day-to-day responsibilities but in terms of their career path as well. Meaning, saying things like "Carol, I really want you to come and join our team based on what you have told me and where you want to go career-wise in 5–10 years. Let me tell you why this would be a good fit for you and how I will help you, as your manager, get there". This sends the message that you are thinking about their career trajectory and will be an advocate.

> From a process standpoint, if a CEO says that a data leadership role is critical to their business, and when a candidate is found who is willing to interview with the company, the company takes a week to set up an interview, you just told this candidate that this is not important to you.—Guy Gomis, Data Science Executive Recruiter

The candidates you want have strong Data Science skills but also good communication skills. They can do the work as well as sit down with stakeholders and understand the problems that they are trying to address. Additionally, they are able to effectively communicate back to the stakeholder and help them change how they behave from a business standpoint. There is a level of emotional intelligence that goes along with it and those are the candidates in highest demand. However, a candidate who has that combination of technical skill and emotional intelligence will pick up on subtle clues in the process that send up red flags regarding how committed the company is to the process of change and this role in particular.

4.3 Reasonable Expectations

In general, across the country, there are regions that have a higher concentration of technical jobs and applicants, specifically locations like New York or the Northern California Bay Area (See Sect. 2.5.4). In those regions, the cycle from posting a data scientist position to hiring a candidate is usually shorter than the national average for junior or mid-level positions. In 2019, the industry average time to fill for Data Science positions was reported at 62 days, some roles had lower numbers, like data scientist with 60 days. Meanwhile, hiring a senior data scientist was taking 70.5 days on average [4].

Also, the Society of Human Resource Management produced a report on "The Global Skills Shortage" reporting that the global talent gap in 2019 was greatest in middle-skilled jobs (Carpentry, plumbing, welding and machining) and high-skilled STEM jobs (Data analysis, science, engineering and medicine). In that report, 83% of HR professionals said they had difficulty recruiting suitable job candidates in the past 12 months and 75% of those said there were skills gaps in job candidates [5].

The challenge is also that different companies will call Data Science different things. A company requesting a data scientist sometimes really describes a business intelligence position where it is a reporting function, almost IT. For this type of role, postings it as a data scientist will mean overpaying or getting someone overqualified for the work. O'Reilly puts out a yearly Data Science salary guide that gives some perspective on salaries, skills and tools used around the country [6]. They usually skew toward candidates with a strong background in developing algorithms or coding, whether it's with tools like R, SQL, or Python.

Also, reasonable expectations should include doing some background work on the evolution of the field as a whole and the relative age of the tools you are seeking. For instance, many companies have standard language in job postings, for a certain salary level, of 10 years of experience in the field.

Several years ago, a few companies put out postings for data scientists with a requirement of 10 years of experience using a new open source tool. At the time, the tool that had only been in existence for 3 years and was already reaching market penetration. Needless to say, they could not find anyone. A larger problem was that even when they realized their mistake they had trouble. Reducing the years of experience down to 3 years automatically reduced the maximum salary they could offer according to company policy. This made them equally non-competitive for finding candidates. They had to lose time in making the case to change existing policies. This is where smaller and more agile organizations can sometimes have an early mover advantage in getting talent.

Therefore, the simple choice of a job title has not been standardized. When a job is listed to hire for a data visualization expert it is not immediately clear what that means in terms of skills, salary, and appropriate years of training. If a hiring manager or an HR department filters for candidates with previous roles as a data visualization expert they may get very few hits, as the field is new.

4.3.1 The Data C-Suite

Over the last several years the importance of data has shown itself in the creation of new C-suite positions in many companies such Chief Data Officer, Chief Analytics Officer, and more. A true C-suite executive will require people management skills, the ability to hire and develop people. As mid-level managers move up to directors and chief analytics officer type roles, they really need to be less hands-on, but knowledgeable enough to know best practices and be credible with the data scientist team. More and more, however, it is the strategic skills that are key. How do we build a team? What skillsets are we looking for? Which tools should be acquired? And lastly, it is the ability to be a change agent. Many companies are just building their data capability, though they have never really had people with these skill-sets at a senior level. Once they decide to make the investment, there will need to be someone who can come in and help the company change how they behave day-to-day.

> At the senior level, the biggest challenge is retention. If you look at people with vice president titles and above in analytics and Data Science, the turnover right now is incredible. People spend less than approximately 18 months in their positions.—Guy Gomis

There appear to be a combination of factors. One obvious factor is the demand for their skills. Candidates, even when they are happy, are being actively recruited. Two, there can be frustration in terms of how companies calibrate the data leadership positions. Senior-level candidates do not always come from the traditional business school backgrounds. They are told they are wanted to bring profits up by suggesting new ideas and acting as change agents. Once hired, however, if a new CDO comes in and tells an established company that the way they are behaving is not working and they give a solution, many times companies are not willing to change. Senior-level candidates in some cases are not willing to stay and fight that battle. They leave because they have other opportunities.

A good example is to imagine a retailer selling apparel who has an expert merchant who decides what to put in the front of the stores. That trusted merchant wants to put blue jeans in the front of every store in the country. The data scientist builds an algorithm that says that blue jeans will sell well in urban America, however, white jeans will sell better in rural America. The typical retail manager would have more trust in the merchant. They have worked in the past and the data scientist just arrived. Also, they do not understand, in an intuitive way, why the algorithm gave that recommendation.

Companies really need to take inventory of whether they are ready to take on a data-driven culture, one that makes decisions based on data. There is an educational and trust-building process that has to happen company wide. If the CEO is really committed to change, the sole responsibility for winning over hearts and minds to the change cannot rest entirely on the shoulders of the Chief Data Hire. Sometimes organizational change companies or consultants are needed to properly change culture and embrace new technological shifts.

4.4 Skills Assessment

Ideally, we need to acknowledge that building a robust Data Science capability requires a team of individuals. The Data Science "unicorn" with all the requite skills is increasing hard to find and even when found poses a long-term risk to organizational stability. Individual skills and personnel, therefore, should be assessed based on the necessary team dynamic not a cookie cutter fit for all hires. Each team member may have gaps in their skill sets, but as long as the team collectively are able to perform when needed it still leaves everyone opportunities for growth.

For a Data Science team, the categories of skills that are vital to assess are outlined below and further described in the section above. (See Sect. 2.3) [7]. Questions to help evaluate candidate's skill level can be extrapolated from Sect. 3.5.1 on Data Science Learning Levels.

The first critical skill is the ability to know how to deal with ambiguous requirements or "problem scoping". Problem scoping is the ability to explicitly plan the analysis, set milestones and stopping rules for each milestone, anticipate and address competing explanations, and determine the best way to evaluate the results.

> If I were given one hour to save the planet, I would spend 59 minutes defining the problem and one minute resolving it.—Albert Einstein

Second, data professionals need to be able to conduct analyses by exploring the data appropriately, building or applying appropriate algorithms, and clearly documenting findings. These skills often receive the bulk of attention when people talk about Data Science. The decisions, however, that data scientists make will depend almost entirely upon the details of the data itself and the tools will vary based on the problems.

Third, is the ability to incorporate analyses into pipelines or smoothly handoff the data. This includes things like reading and writing data to/from any format and location, incorporating complex matching and filtering, as well as making work compatible with the engineering stack.

Fourth is to incorporate pipelines into the business, which generally translates into "communication skills". This can be assessed with some standard business tools. A data scientist often serves as a bridge between technical and non-technical stakeholders, and so should be able to understand the non-technical side of the business. The ability to navigate the business's organizational structure including relationship building and stakeholder management is key. Packaging technical work for diverse audiences is an art which includes non-technical communication and visualization for effect as well as information. Have the candidate present a technical topic for a technical audience and a non-technical one making clear that they should tailor their presentation to each.

Finally, engagement with the larger profession is ideal. It flags an individual contributor as someone whose ability to contribute meaningfully has been vetted. Also, because Data Science as a profession is changing too rapidly for a Data Science team to really keep up with the state of the art all team members must continuously remain current with how the field is evolving and what changes are most appropriate for the needs of the business.

With the support of leadership, this process can be instilled in the team culture by the team lead by participating in external Meetups, hackathons, open source code repositories, blogs, and other public contributions to keep the team connected with the wider community. Public contributions invite public comment and criticism, which in turn improves practice.

4.5 Talent Sourcing: An Eye Toward Diversity

There are many sources of data and analytics talent to consider and explore. An eye toward diversity will help smart practical managers in sourcing talent effectively. The popular portrayal of a data or technology person is a young, single, male, recent college graduate who may be socially awkward or a loner. The question, however, is whether this vision traps employers into certain recruitment tracks.The data field is young, but does that mean that young people have a better handle on it? Communication skills and end-to-end thinking may be missing in a recent graduate with no prior work experience. Also, the vision of the socially awkward loner can be challenged as fundamentally not the right skill set for a data scientist who may need to build bridges between business units or stakeholders to get things done. Every part of this vision can unknowingly create talent blind spots that can be evaluated by your team for innovative solutions. For example, recruiting at Meetups, college campuses, and informal weekend tech events only, unconsciously selects for young, single people with the time to spend their evenings and weekends in this way. Also recruiting only at top-tier universities narrows your scope and greatly increases your competition. Recruiting high achievers from mid-tier schools, community colleges, and re-training focused bootcamps can produce strong dividends particularly when combined with offering training opportunities. D.J Patel the former Chief Data Scientist for the United States credits his community college background as part of his overall success. Also supporting data science events at forums that host large numbers of women and minority engineers and scientists, such as HBCU's, professional societies, and non-traditional programs, can pay off in large measure to increase your available pool of high-quality talent. All options are viable, but an eye toward diversity can greatly expand on traditional options to achieve a competitive advantage.

4.5.1 *Academia*

By partnering with academia as a training mechanism, industry can identify gaps in the current Data Science and Analytics landscape that warrant the greatest attention from training providers. The can also work to create workforce development pipelines for bringing in new talent as well as "up-skilling" existing workforce. Important facets of a successful academic-industry partnership include a collaborative rather than a transactional relationship, coupled with alignment of motivations, and timelines.

A complexity of finding Data Science and Analytics talent at universities is that data skills are scattered across campus in different departments, institutes, and centers. Only recently have dedicated programs begun to emerge at institutions and often they may not encompass all of the talent available on campus. Some universities have taken up this challenge by creating Data Science institutes that have an industry affiliates program that companies can join. This usually is an indication of at least an awareness of the problem and an avenue for companies to engage. There are direct academic corporate relations functions, not to be confused with corporate relations through the entire university advancement office. Different universities have different titles, but at most there is a central corporate relations office's whose mission is to receive gifts for university advancement. Make sure the advancement office understands the unique challenges of Data Science and will work to ensure due diligence once the relationship begins.

Data Science Focused Institutes
Data science focused institutes can work as insider advocates for companies within the university structure. Schools that host "Industry Days or Demo Days" can give a better indication of the level of industry collaboration already in place, along with an indication of the institution's connections with companies.

Other ways to engage include looking for research being done in an area of interest and contacting those publishing in your area. Engage in challenges, hack-a-thons, and competitions to source talent as well as project-based courses to get students working on problems in your area. As stated in the beginning, this level of engagement requires, a collaborative rather than a transactional relationship with the university or college, often involving resources in the form of affiliate program dues, student prize money, event sponsorship or all of the above. Consider allotting personnel time to speak, judge, or participate in classes. Universities know now that students are their biggest commodity with industry, particularly in the Data Science area, so they are becoming savvier. Many are not offering as many free student interaction opportunities as in the past or restricting these to less competitive areas. Because talent is scarce some universities reserve the most rewarding opportunities for paying partners. This means being prepared for higher price tags at top tier universities to even engage with students.

Committing to deep engagement with academic partners can yield multiple rewards, increased access to the best and brightest students and partnering with faculty doing research. Industry often prioritize institutions that will promote business cases or accelerate the adoption of Data Science outcomes/deliverables that lead to a return on the investment for the company. However, companies must be mindful that requiring strict intellectual property control and short (less than 1 year) deliverables may conflict with the culture at most universities, present contractual challenges and seriously hinder the adoption they might hope for with faculty or administrators. In part, this is because academics are not usually rewarded for their work with industry. It is frequently seen as a mark against them for tenure and promotion within some departments. Companies must therefore be cognizant of the fact that unless they are engaging with a university Data Science institute whose purpose is Data Science recruitment, an individual faculty member is not motivated to engage unless there is a clear mutual interest in a topic. Funding alone, unless offered in the millions, will not be able to sway faculty to pivot to topics in their research or classes that do not build toward promotion or will require major changes to curriculum, significant time, or levels of approval. Most of the revenue for these institutions come from federal research dollars and faculty tend to focus on these efforts leaving little time to tailor industry engagement that strays too far away from basic research support.

If your organization is established in the analytics field and is looking to push the boundary of what is possible within advanced algorithms, research institutions offer world class groups focused on experimental development that can work with your engineers or research groups. Major research universities, however, tend to have a higher price tag for industry affiliates programs and can be less flexible to any one partner's needs. Depending on the company needs, the higher price tag and overall competitive nature of large universities may be prohibitive. An alternative can be forming a proactive relationship with smaller schools, 4-year colleges, community colleges, or trusted technical college programs.

Primarily undergraduate institutions may be a good place to engage in advertising internships and participating in the education process. They offer opportunities to guest lecture and to expose students to the field. Project based courses are excellent ways of requesting that students work with your employees to solve industry problems, instead of simply providing them an answer. This can benefit both parties by allowing a 2-way exchange of exposure to new Data Science methods and business problems. Additionally, at the two year and technical college level, there may be more students in the pipeline for data-enabled degrees, as they can pivot to vocational training [8] even if the current focus is on fundamental skills needed for direct transfer to 4-year colleges [9].

To build a workforce pipeline of data and analytics talent that can tackle a broad array of problems in different industries, educators are needed who can stage learning, and helping students stack skills (Fig. 3.1). If this stacking is not achieved, the consequences include higher costs for organizations as they struggle to find adequate talent for the roles that are most in demand, and more potential turnover as workers move between firms to gain the on the job experience they need to be highly valued. Therefore, long-term investment in educational institutions should be a core value in the best interest of every company or hiring team.

4.5.2 *Startups*

Partnering with a local startup incubator provides an excellent way to collaborate, especially since we are now seeing a startup economy that is more geographically diverse than ever before [10]. Startups are also a source of innovation and proof of concept. Several large companies have been acquiring startups not only for their products, intellectual property, and brand, but also for their workforce. The term "acqui-hires" has exploded as a way of saying companies that were purchased primarily for recruiting; where the whole team is hired into the acquiring organization but the product or service they were bringing to market may not be adopted. This is being used as a major resource for large companies in the position to acquire talent. One issue to remain aware of is culture fit. If the culture of the large company is substantially different from the ethos of the start-up there may be challenges in retaining the new-found talent. This can be overcome in some cases by clear communication and goals, allowing a measure of autonomy for the new group as they merge with existing teams.

4.5.3 Industry Partners, Consultants, and Non-Traditional Partners

Joining sector professional societies may be another way for participating organizations to come together to establish industry-recognized credentials, such as badges, that would satisfy work requirements for students who undertake work-based learning [11]. Skills determination as discussed above is a daunting task. This is why Data Science hiring consulting firms such as Data Driven Vision, Burtchworks, or Brainworks have emerged to assist in hiring strategies.

Non-traditional partners, bootcamps or specialized training programs, add to the landscape as well. For instance, some well-established programs like Per Scholas, founded more than 20 years ago, will design courses for entry-level workers in response to employer needs, including in Data Science and analytics. The biggest pros of this model are freedom for employers to customize change and adapt, as well as tap into new communities for talent. This path does require a hands-on approach and knowing more fully what you want from employees in the short term. In rare cases, this may need to be monitored as a long-term strategy, as some of these institutions do not have the track record of longevity and development work may be lost if they go out of business.

Finally, a social component can be used very effectively. Sending employees out to groups and other social gathering spots to search for candidates has a lasting effect born out of a culture of recruiting within an organization (See Sect. 4.2). Participating in public open competitions such as Kaggle and others can help identify teams of individuals with good talent.

4.6 Continuing Education for Managers and Workers

One highly important fact for career seekers as well as hiring managers and professionals is that a career in Data Science and analytics is ever evolving and the pace of that evolution is also ever increasing. This should give hope to new entrants. There is always a space in which you can be an early adopter.

What will serve institutions well is investing in the infrastructure for continuous learning. Putting together practical steps to learn new things at regular intervals and a process for expanding and communicating that knowledge effectively to others is paramount. Investing first in innovation skills like communication, translation of data topics into use cases, framing a problem around data, data visualization and end-to-end thinking will enable proper identification of goals and the ability to move forward.

Once these innovation skills have been mastered, organizations are poised to learn any number of new skills and apply them to their core business. Curiosity is key. Once a culture of curiosity is developed new capabilities can be learned as needed in a timely and cost-effective manner. We discuss the most common ways in the Sect. 2.3.

Degree programs are becoming more flexible in their hours for working professionals. Hands-on coursework with projects validating practical skills are increasingly accepted by promotion committees and hiring managers in lieu of a formal degree in Data Science. Data science certificates are a great way to gain experience and knowledge in Data Science without the time and expense of earning a degree. Bootcamps and free online courses also fit the bill if internal workers need only to demonstrate proficiency in applying data to a use case.

Technologically mature companies may want to design in-house Data Science training programs or offer incentives or reimbursements for employees to get external training. Data science training for employees could include mentorship programs or groups, internal Data Science bootcamps and hack-a-thons, idea pitch programs, and consultant run workshops.

There is no "one size fits all". Programs can be offered online or in person (or as a hybrid). Also, community colleges serve as an access point to permit existing members of the workforce to retrain or obtain specific new skill sets to complement their education and experience. Any way that it is delivered, data skills, training, and professional development will pave the way for innovation and success.

References

1. Burtch, L. (2017). *The Burtch Works Study: Salaries of Data Scientists*. Burtch Works, Executive Recruiting. http://www.burtchworks.com/wp-content/uploads/2017/05/DS-2017-Industry.pdf
2. BrainWorks. (2018, December 11). Best hiring practices: How the winners of the recruitment game are playing [Blog post]. Retrieved from https://brainworksinc.com/best-hiring-practices-how-the-winners-of-the-recruitment-game-are-playing/
3. AIRS-ADP (2019) Professional Recruiter Certification 2.0 Retrieved from https://www.airsdirectory.com/mc/airs_prc_20_adp.dbprop?_mhid=4888117461
4. Howden, D. (2019, April 19). *What is time to fill? KPIs for recruiters*. Retrieved from https://resources.workable.com/blog/recruiting-kpis
5. (2019). *The global skills shortage: Bridging the talent gap with education, training and sourcing*. Alexandria, VA. SHRM
6. Suda, B. (2018). *2017 Data Science Salary Survey*. O'Reilly Media, Inc.
7. Wheeler, S. (2018, February 27). *A framework for evaluating data scientist competency*. Retrieved from https://towardsdatascience.com/a-framework-for-evaluating-data-scientist-competency-89b5f275a6bf
8. Rawlings-Goss. (2018). *Keeping Data Science Broad: Negotiating the Digital & Data Divide*. National Science Foundation. Retrieved from https://drive.google.com/file/d/14l_PGq4AxOP9fhJbKqA2necsJZ-gdiKV/view
9. Gould, R., Peck, R., Hanson, J., Horton, N., Kotz, B., Kubo, K., Malyn-Smith, J., Rudis, M., Thompson, B., Ward, M., Wong, R. (2018). *The two-year college data science summit Report*. https://www.amstat.org/asa/files/pdfs/2018TYCDS-Final-Report.pdf
10. Mandel, M. (2017, March 30). *How the startup economy is spreading across the country – and how it can be accelerated*. Retrieved https://www.progressivepolicy.org/issues/economy/how-the-startup-economy-is-spreading-across-the-country/
11. Business-Higher Education Forum. (2018). *Building a diverse cybersecurity talent ecosystem to address national security needs*. http://www.bhef.com/sites/default/files/2018BHEFUSMCaseStudy.pdf

Chapter 5
Conclusion

To lay the foundation for success in forming a data career, degree program or business team we must return to a point from the introduction. Data is not where the true power is; it is in people. Not just the people who code but the people who use data to make decisions and those whose lives data effects.

For good or for ill, it will be people who manipulate data to gain insights and power over how our world is shaped. It will be people who form our perception of data as well. Therefore, when data scientists write algorithms we must consciously decide, as in real life, whether we are representing life as it is or striving toward a higher goal.

These things are true because, unfortunately, data cannot take the place of empathy and on its own it is not objective, unbiased, or clean, unless we force it to be so. Perfect data can only reflect back the world as it is, and project it into the future with all of its hang-ups, unfairness, and bias. It is merely a powerful augmentation to our own thought process. It is an assistant, that we should highly regard, in order to make sure our personal blind spots do not keep us from seeing a holistic view. This is the true power of data.

People with keen minds are needed. Having a diverse set of people in the pipeline helps to vet the data, making sure it is representative of a broad number of views before it is accepted as ground truth about the world. You need a diverse set of people to ask the question "Who or what is being left out of the data?". Consider, Twitter data, which seems large, but it could leave out the elderly or those without computer access. Are we taking data about a subset of individuals and assuming it represents everyone? Where can data be unprofitable, harmful, or even dangerous? For instance, hospital outbreak data could leave out rural communities that do not have major hospitals nearby to record their cases. In this vein, I recommend another book "Weapons of Math Destruction" by Cathy O'Neil, where the point is beautifully laid out. It illustrates that how you use data matters and using numbers does not give us objectivity or impartiality.

R. Rawlings-Goss, *Data Science Careers, Training, and Hiring*, SpringerBriefs in Computer Science, https://doi.org/10.1007/978-3-030-22407-3_5

The ability to help achieve and crystalize diversity of thought is what is pushing us into the future. Beyond the initial bursts of revelation, that can come from seeing insights with data. People are needed to push the boundaries beyond our current structures. Radically new ideas will not have any data to back them up. They will have to be taken, as always, as a leap of faith.

5.1 The New World

In the introduction, people who chose data careers are compared to doctors. There is a high amount of impact, a diversity of specialties, and a large number of problems to diagnosis. So, like doctors, we also need to make a formal commitment to "do no harm".

The new world tells us that data is very powerful, so more than ever we want to take it a step further to commit to *doing* Good with Data. This will include the entire ecosystem of data professionals, students, degree programs, companies, and people of all types being engaged, including women and people of color.

Getting Good with Data requires skills, knowledge, and the ability to assess what and where to go next.

Doing Good with Data is a requirement to view "the world of data" through the eyes of "the people of the world".

This shows us that "Data Does Not Have the Power to Shape our Future". We Do! It is *us*!

Chapter 6
Resources

6.1 Program Overview: Over 450 Data Science Degree Programs Across the Nation

Table 6.1 A collection of 468 Data Science and related programs, including Bachelors, Masters, Doctoral, Associates, and Certificate programs

School	Program	Degree
Albright College	M.S. in Business Intelligence	M
American Sentinel University	Master of Science Business Intelligence and Analytics (MSBIA)	M
American University	MS Business Analytics	M
American University	Online MS Analytics	M
Arizona State University	Advanced Analytics in Higher Education	C
Arizona State University	Business Analytics	M
Arizona State University	Business Data Analytics	B
Arizona State University	Master of Science in Business Analytics	M
Arkansas Tech University	Business Data Analytics	B
Aspen University	Master of Science in Technology and Innovation—Business Intelligence and Data Management	M
Auburn University	Data Science	B
Auburn University	Online Master of Business Administration with concentration in Business Analytics	M
Aurora University	Master of Science in Digital Marketing and Analytics	M
Austin Peay State University	Professional Science Master's Degree in Predictive Analytics	M
Austin Peay State University	Professional Science Master's in Data Management and Analysis	M
Babson College	MBA with Business Analytics Concentration	M
Baker College	MBA in Business Intelligence	M

R. Rawlings-Goss, *Data Science Careers, Training, and Hiring*, SpringerBriefs in Computer Science, https://doi.org/10.1007/978-3-030-22407-3_6

School	Program	Degree
Becker College	Data Science	B
Bellevue College	Healthcare Data Analyst Certificate	C
Bellevue University	Master of Professional Science in Technology Innovation with Focus in Bioinformatics	M
Bellevue University	Master of Science-Business Analytics	M
Benedictine University	Master of Science in Business Analytics	M
Bentley University	Graduate Certificate in Business Analytics	C
Bentley University	Master of Business Analytics	M
Bentley University	Master of Science in Marketing Analytics	M
Boston University	Graduate Certificate in Data Analytics	C
Boston University	Graduate Certificate in Database Management and Business Intelligence	C
Boston University	Master of Science in Computer Information Systems Online (Data Analytics Concentration)	M
Boston University	Master of Science in Computer Science (Data Analytics Concentration)	M
Bowling Green State University	M.S. in Analytics	M
Bowling Green State University	Master of Science in Applied Statistics (specialization in Business Analytics)	M
Brandeis University	M.S. in Strategic Analytics	M
Brandeis University	Master of Arts in Computational Linguistics	M
Brigham Young University—Idaho	Data Science	A
Brigham Young University—Idaho	Data Science	B
Brigham Young University—Idaho	Data Science	C
Brown University	Big Data	D
Bunker Hill Community College	Data Management (Fast-Track) Certificate Program	C
California Polytechnic State University	Business Analytics	M
California Polytechnic State University	Data Science Minor	B
California State University-East Bay	Master of Science in Business Administration: Business Analytics Option	M
California State University-Fullerton	Certificate in Data Science	C
California State University-Fullerton	Certificate in Healthcare Analytics	C
California State University-Fullerton	Master of Science in Information Systems and Decision Sciences	M
California State University-Fullerton	MBA with Business Analytics	M

School	Program	Degree
California State University-Long Beach	Master of Science in Applied Statistics	M
California State University-San Bernardino	Master of Business Administration (M.B.A.)—Business Intelligence and Information Technology Focus	M
Capella University	Business Intelligence Specialization Post-Baccalaureate Certificate	C
Capella University	MBA in Business Intelligence	M
Capella University	MS in Analytics	M
Carnegie Mellon University	Machine Learning	D
Carnegie Mellon University	Master of Computational Data Science (MCDS)	M
Carnegie Mellon University	Master of Information Systems Management, Business Intelligence and Data Analytics (MISM-BIDA)	M
Carnegie Mellon University	Master of Science in Information Technology, Business Intelligence and Data Analytics (MSIT-BIDA)	M
Carnegie Mellon University	Master of Statistical Practice	M
Carnegie Mellon University	MBA: Business Analytics Track	M
Case Western Reserve University	Data Science	B
Case Western Reserve University	Health Informatics	M
Case Western Reserve University	Systems Biology and Informatics	D
Central Connecticut State University	Data Mining	M
Central Connecticut State University	Graduate Certificate in Data Mining	C
Central Michigan University	Applied Statistics and Analytics	M
Central Michigan University	Data Mining	C
Central Piedmont Community College	Data Analytics	A
Cerro Coso Community College	Data Analyst I Certificate	C
Chapman University	Computational and Data Sciences	D
Chapman University	Master of Science in Computational and Data Sciences	M
City University of Seattle	Master of Science in Computer Systems with Concentration in Big Data Management	M
Claremont Graduate University	Masters of Science in Information Systems and Technology: Concentration in Data Science and Analytics	M
Cleveland State University	Graduate Certificate in Advanced Business Analytics	C
Cleveland State University	Graduate Certificate in Strategic Business Analytics	C
College of Charleston	Data Science	B

School	Program	Degree
Colorado State University-Fort Collins	Applied Statistics Certificate	C
Colorado State University-Fort Collins	Data Analysis Certificate	C
Colorado State University-Fort Collins	Data Science	M
Colorado State University-Fort Collins	Graduate Certificate in Business Intelligence	C
Colorado State University-Fort Collins	Master of Applied Statistics (M.A.S.)	M
Colorado State University-Global Campus	Certificate in Business Intelligence	C
Colorado Technical University	Big Data Analytics	D
Columbia University in the City of New York	Master of Arts in Biomedical Informatics	M
Columbia University in the City of New York	Master of Arts in Quantitative Methods for Social Sciences	M
Columbia University in the City of New York	Master of Arts in Statistics	M
Columbia University in the City of New York	Master of Science in Applied Analytics	M
Columbia University in the City of New York	Master of Science in Computer Science with concentration in Machine Learning	M
Columbia University in the City of New York	Master of Science in Data Journalism	M
Columbia University in the City of New York	Master of Science in Data Science	M
Columbia University in the City of New York	Masters in Applied Statistics	M
Columbia University in the City of New York	PhD in Biomedical Informatics	M
Columbia University in the City of New York	PhD in Statistics	D
Community College of Allegheny County	Data Analytics Technology Associate of Science	A
Cornell University	Data Analytics	M
Cornell University	Master of Professional Studies (MPS) in Applied Statistics (Option II: Data Science)	M
Cornell University	Master of Science in Health Informatics	M
Cornell University	MBA	M
Creighton University	Master of Science in Business Intelligence and Analytics (MS-BIA)	M
Creighton University	Master of Science in Data Science	M
CUNY Bernard M Baruch College	Advanced Certificate in Business Analytics of Large-Scale Data	C

School	Program	Degree
CUNY Graduate School and University Center	Online Master of Science in Data Analytics	M
CUNY Queens College	MA in Data Analytics and Applied Social Research	M
Dakota State University	Analytics	M
Davenport University	Master of Science in Health Informatics and Information Management	M
Denison University	Major in Data Analytics	B
DePaul University	Information Systems (MS)—Business Intelligence Concentration	M
DePaul University	M.S. in Predictive Analytics	M
Drexel University	Certificate in Healthcare Informatics	C
Drexel University	Information Studies	D
Drexel University	M.S. in Business Analytics	M
Drexel University	Online Master's in Health Informatics	M
Duke University	Master in Interdisciplinary Data Science	M
Eastern Michigan University	Enterprise Business Intelligence specialization in the Master of Business Administration (MBA)	M
Eastern Michigan University	Master of Arts in Applied Statistics	M
Elmhurst College	Graduate Certificate in Data Science	C
Elmhurst College	M.S. in Data Science	M
Elon University	Information Science	B
Florida International University	Master of Science in Health Informatics and Management Systems	M
Florida Polytechnic University	Big Data Analytics	B
Fordham University	Master of Science in Business Analytics	M
Full Sail University	Master of Science in Business Intelligence	M
Galvanize U	Master of Engineering in Big Data	M
George Mason University	Computational Data Sciences Minor	B
George Mason University	Computational Sciences and Informatics	D
George Mason University	Masters in Information Systems	M
George Mason University	MS in Computational Science	M
George Mason University	MS in Data Analytics Engineering	M
George Mason University	Spatial Business Intelligence Graduate Certificate	C
George Washington University	Master of Science in Data Science	M
Georgetown University	Master in Data Science for Public Policy (MDSPP)	M
Georgetown University	Master of Science in Analytics, Concentration in Data Sciences (MS-DS)	M
Georgia Southern University	M.S.—Computer Science with concentration in Data and Knowledge Systems	M
Georgia State University	Master of Science in Analytics	M
Georgia Tech	Machine Learning Ph.D	D

School	Program	Degree
Georgia Tech	MS—Computer Science with specialization in machine learning	M
Georgia Tech	MS in Analytics	M
Grand Valley State University	Master of Science in Biostatistics	M
Grantham University	Master of Science in Business Intelligence	M
Great Bay Community College	Certificate in Practical Data Science	C
Harrisburg University of Science and Technology	Online Master of Science in Analytics	M
Harvard University	Data Science Certificate	C
Harvard University	Master of Science in Computational Biology and Quantitative Genetics	M
Harvard University	Master of Science in Computational Science and Engineering	M
Illinois Institute of Technology	Data Analytics	M
Illinois Institute of Technology	Master of Data Science	M
Illinois Institute of Technology	MS—CS specialization in Data Analytics (with co-terminal BS/MS option)	M
Indiana University Bloomington	Business Analytics Certificate Program	C
Indiana University Bloomington	Informatics	D
Indiana University Bloomington	Master of Science in Data Science	M
Indiana University Bloomington	MBA (Major in Business Analytics)	M
Indiana University Bloomington	Online Certificate in Data Science (Graduate)	C
Indiana University Bloomington	Online MS in Business Analytics	M
Indiana University-Purdue University-Indianapolis	M.S. in Mathematics: Applied Statistics	M
Indiana University-Purdue University-Indianapolis	Master of Science in Bioinformatics	M
Jackson State University	Computational and Data-Enabled Science and Engineering	D
Johns Hopkins University	Bioinformatics	M
Johns Hopkins University	Master of Science in Geographic Information Systems	M
Johns Hopkins University	Master of Science in Information Systems	M
Johnson County Community College	Data Science, Data Analytics Certificate	C
Keller Graduate School of Management	MBA with a Business Intelligence and Analytics Management Concentration	M
Kennesaw State University	Analytics and Data Science	D
Kennesaw State University	Master of Science in Applied Statistics	M

School	Program	Degree
Kennesaw State University	Online Certificate in Applied Statistics	C
La Salle University	M.S. in Analytics	M
Lewis University	Business Analytics, M.S.	M
Lewis University	Data Science	M
Loras College	MBA—Business Analytics	M
Louisiana State University	Master of Science in Analytics	M
Loyola University Chicago	Graduate Certificate in Business Intelligence and Data Warehousing	C
Loyola University Chicago	M.S. in Applied Statistics	M
Luther College	Data Science	B
Manchester Community College	Applied Data Analytics Certificate	C
Marist College	Advanced Certificate in Business Analytics	C
Maryville University	M.S. in Business Data Analytics	M
Medical University of South Carolina	Master of Science in Health Informatics	M
Mercer University	M.S. in Business Analytics	M
Mercyhurst University	Data Science	M
Miami Dade	Business Intelligence Professional College Credit Certificate	C
Miami Dade	Business Intelligence Associate Degree	A
Miami University of Ohio	Business Analytics Minor	B
Miami University of Ohio	I.S. and Analytics	B
Michigan State University	M.S.—Business Analytics	M
Michigan State University	Master of Science in Applied Statistics	M
Michigan Technological University	Master of Science Degree in Integrated Geospatial Technology	M
Middle Tennessee State University	Master of Science in Information Systems (Business Intelligence and Analytics Concentration)	M
Middle Tennessee State University	Master of Science in Professional Science (M.S.) with a concentration in Health Care Informatics	M
Misericordia University	Health Care Informatics Graduate Certificate	C
Misericordia University	Health Informatics Master's Program	M
Missouri University of Science and Technology	Graduate Certificate in Business Analytics and Data Science	C
Nashua Community College	Foundations in Data Analytics	A
National University	Master of Science in Data Analytics	M
National University	Master of Science in Health and Life Science Analytics	M
Nebraska College of Technical Agriculture	Business Analytics Graduate Certificate	C
New College of Florida	Master of Data Science	M
New Jersey Institute of Technology	Graduate Certificate in Data Mining	C
New Jersey Institute of Technology	MS in Applied Statistics	M

School	Program	Degree
New York University	Advanced Certificate in Applied Urban Science and Informatics	C
New York University	Applied Data Analytics and Visualization	B
New York University	Certificate in Healthcare Informatics	C
New York University	Master of Science in Applied Statistics for Social Science Research	M
New York University	Master of Science in Business Analytics	M
New York University	Master of Science in Data Science	M
New York University	Master of Science in Information Systems	M
New York University	MBA, Specialization in Business Analytics	M
New York University	MS in Applied Urban Science and Informatics	M
New York University	PhD in Data Science	D
Normandale Community College	Data Management and Analysis	A
North Carolina State University at Raleigh	Master of Science in Analytics	M
Northeastern University	Graduate Certificate in Data Analytics	C
Northeastern University	Graduate Certificate in Game Analytics	C
Northeastern University	Graduate Certificate in Learning Analytics	C
Northeastern University	Graduate Certificate in Urban Analytics	C
Northeastern University	Master of Professional Studies in Analytics	M
Northeastern University	Master of Professional Studies in Informatics	M
Northeastern University	Master of Science in Business Analytics	M
Northeastern University	Master of Science in Data Science	M
Northeastern University	Master of Science in Health Data Analytics	M
Northeastern University	Master of Science in Health Informatics	M
Northeastern University	Master of Science in Urban Informatics	M
Northern Kentucky University	Data Science	B
Northwestern College	Analytics Certificate	C
Northwestern University	Certificate in Analytics and Business Intelligence for IT Professionals	C
Northwestern University	Certificate in Predictive Business Analytics	C
Northwestern University	Master of Science in Analytics	M
Northwestern University	Master of Science in Information Systems with Concentration in Analytics and Business Intelligence	M
Northwestern University	Master of Science in Predictive Analytics Online	M
Notre Dame College	Certificate in Competitive Business Intelligence	C
Notre Dame of Maryland University	Master of Science in Analytics in Knowledge Management	M
Nova Southeastern University	Graduate Certificate in Business Intelligence/Analytics	C
Nova Southeastern University	Master of Science in Biomedical Informatics	M

School	Program	Degree
Oakland University	Master of Science in Information Technology Management in Business Analytics	M
Oakland University	MS in Applied Statistics	M
Ohio University	Online MBA in Business Analytics	M
Oklahoma State University Center for Health Sciences	Graduate Certificate in Business Data Mining	C
Oklahoma State University Center for Health Sciences	Graduate Certificate in Marketing Analytics	C
Oregon State University	MBA in Business Analytics	M
Pace University-New York	Master of Science in Information Systems	M
Pace University-New York	MS in Customer Intelligence and Analytics	M
Pasco-Hernando State College	Healthcare Informatics Specialist Certificate	C
Pennsylvania State University	Data Sciences	B
Pennsylvania State University	Graduate Certificate in Applied Statistics	C
Pennsylvania State University	Graduate Certificate in Business Analytics	C
Pennsylvania State University	Graduate Certificate in Business Analytics	C
Pennsylvania State University	Master of Applied Statistics	M
Pennsylvania State University	Master of Professional Studies in Data Analytics	M
Pennsylvania State University	Master of Professional Studies in Data Analytics	M
Pennsylvania State University	Social Data Analytics	D
Philadelphia University	M.S. in Modeling, Simulation and Data Analytics	M
Purdue University-Main Campus	Certificate in Applied Statistics	C
Purdue University-Main Campus	MBA, Specialization in Business Analytics	M
Quinnipiac University	Master of Science in Business Analytics	M
Radford University	MS in Data and Information Management	M
Regis University	Data Science	M
Regis University	Graduate Certificate in Data Science	C
Rensselaer Polytechnic Institute	M.S. in Information Technology—Concentration in Data Science and Analytics	M
Rensselaer Polytechnic Institute	MS in Business Analytics	M
Rockhurst University	Business Intelligence and Analytics	M
Rockhurst University	Data Science and Business Analytics Certificate	C
Rutgers University	Master of Business and Science Degree in Analytics—Discovery Informatics and Data Sciences	M

School	Program	Degree
Rutgers University	MBA with Analytics and Information Management Concentration	M
Rutgers University	Online Master of Information	M
Rutgers University	Post MBA Certificate in Analytics and Information Management	C
Saint Joseph's University	Master of Science in Business Intelligence and Analytics Program	M
Saint Joseph's University	Online Master of Science in Business Intelligence and Analytics	M
Saint Joseph's University	Online Master's in Health Administration: Informatics Specialization	M
Saint Louis University	Health Data Science	M
Saint Louis University-Main Campus	Applied Analytics Master's Degree	M
Saint Mary's College	Master of Science in Data Science	M
Saint Peter's University	Master of Science in Data Science with a concentration in Business Analytics	M
San Jose State University	Online Business Analytics Certificate	C
Santa Clara University	MBA with concentration in Data Science and Business Analytics	M
Seattle University	Graduate Certificate in Business Analytics	C
Seattle University	Master of Science in Business Analytics	M
Sinclair College	Data Analytics DA.S.CRT	C
Smith College	Statistical and Data Sciences	B
South Dakota State University	Computational Science and Statistics	D
South Dakota State University	Data Science	M
Southern Methodist University	Online Master of Science in Data Science	M
Southern Methodist University	Statistics and Data Analytics	M
Southern New Hampshire University	Data Analytics	B
Southern New Hampshire University	Master of Science in Data Analytics	M
Southern New Hampshire University	MBA in Business Intelligence	M
Southern New Hampshire University	MS in Information Technology—Healthcare Informatics	M
Southwestern Oklahoma State University	M.S. Healthcare Informatics and Information Management	M
St. John's University-New York	Master of Science in Data Mining and Predictive Analytics	M
St. Mary's University	Data Analytics	B
Stanford University	Biomedical Informatics MS Degree	M

School	Program	Degree
Stanford University	Data Mining and Applications Graduate Certificate	C
Stanford University	M.S. in Statistics: Data Science	M
Stanford University	Master of Science in Computer Science, Specialization in Information Management and Analytics	M
Stanford University	Mining Massive Data Sets Graduate Certificate	C
Statistics.com	Analytics for Data Science	C
Stevens Institute of Technology	BI&A Graduate Certificate	C
Stevens Institute of Technology	Master of Science in Information Systems	M
Stevens Institute of Technology	MS in Business Intelligence and Analytics	M
Syracuse University	Certificate of Advanced Study in Data Science	C
Syracuse University	Data Science	M
Tarleton State University	M.S. in Mathematical Data Mining	M
Temple University	Master of Science in Health Informatics	M
Tennessee Technological University	Professional Science Master's Degree in Environmental Informatics	M
Texas A & M University-College Station	M.S. in Analytics	M
Texas A & M University-College Station	Online M.S. in Applied Statistics	M
Texas Tech University	Master of Science in Data Science	M
The College of Saint Scholastica	M.S. Health Informatics	M
The George Washington University	Business Analytics	M
The New School	Data Visualization (MS)	M
The Ohio State University	Data Analytics	B
The Ohio State University	Master of Applied Statistics	M
The Ohio State University	Master of Public Health in Biomedical Informatics	M
The University of Alabama	M.S. degree in Applied Statistics, Data Mining Track	M
The University of Alabama	Master of Science in Marketing, Specialization in Marketing Analytics	M
The University of Alabama	MBA with concentration in Business Analytics	M
The University of Alabama	MS in Operations Management—Decision Analytics Track	M
The University of Iowa	Business Analytics	B
The University of Iowa	Business Analytics Certificate Program	C
The University of Iowa	Business Analytics Graduate Program	M
The University of Tennessee	Master of Science in Business Analytics	M
The University of Tennessee	MBA with Business Analytics Concentration	M
The University of Tennessee at Chattanooga	Professional MBA with Business Analytics	M

School	Program	Degree
The University of Texas at Austin	Master of Science in Business Analytics	M
The University of Texas at Dallas	Master of Science in Business Analytics	M
The University of Texas at Dallas	Master of Science in Healthcare Management with Specialization in Healthcare informatics	M
The University of Texas at Dallas	Master of Science in Marketing with Specialization in Marketing Analytics	M
The University of Texas at Dallas	Social Data Analytics and Research	M
The University of Texas at San Antonio	Master of Science in Applied Statistics	M
Thomas Edison State University	Data Science and Analytics	B
Thomas Edison State University	MBA in Data Analytics	M
Trocaire College	Certificate in Healthcare Informatics	C
Union Graduate College	Master of Science in Healthcare Data Analytics	M
University at Buffalo	Master of Science (MS) in Management Science: Business Analytics and Systems from the State University of New York at Buffalo and a Master of Business	M
University of Akron	Data Science	M
University of Arizona	Business Intelligence and Analytics (Certificate NDP)	C
University of Arkansas	Graduate Certificate in Business Analytics	C
University of Arkansas	Master of Information Systems with Business Analytics Concentration	M
University of Arkansas	Professional Master of Information Systems	M
University of Arkansas at Little Rock	Information Quality Program	M
University of Arkansas at Little Rock	Information Quality Program PhD	D
University of California Hastings College of Law	Master of Information Management and Systems	M
University of California-Berkeley	Business Intelligence and SAS Analytics Software	C
University of California-Berkeley	Data Science	B
University of California-Berkeley	Data Science and Systems Concentration	M
University of California-Berkeley	Information and Data Science	M
University of California-Davis	Hybrid MBA with Business Analytics and Technologies Concentration	M
University of California-Davis	Master's of Health Informatics	M
University of California-Irvine	Business Intelligence and Data Warehousing Certificate	C

School	Program	Degree
University of California-Irvine	Data Science	B
University of California-Irvine	Predictive Analytics Certificate Program	C
University of California-San Diego	Data Mining Certificate	C
University of California-San Diego	Master of Advanced Study in Data Science and Engineering	M
University of California-San Diego	Master of Science in Business Analytics	M
University of Central Florida	M.S. in Statistical Computing Data Mining Track	M
University of Central Florida	Master of Science in Health Care Informatics	M
University of Central Florida	Professional Science Master's Program in Health Care Informatics	M
University of Chicago	Master of Science in Analytics	M
University of Chicago	Master of Science in Computational Analysis and Public Policy (CAPP)	M
University of Cincinnati-Main Campus	Master of Science in Business Analytics	M
University of Cincinnati-Main Campus	Master of Science in Health Informatics	M
University of Colorado Boulder	Master of Science in Business Analytics	M
University of Colorado Denver	Graduate Applied Statistics Certificate	C
University of Colorado Denver	Master of Science in Information Systems	M
University of Colorado Denver	MS in Business Analytics—Big Data Specialization	M
University of Colorado Denver	MS in Decision Sciences	M
University of Colorado Denver	MS in Information Systems—Business Intelligence	M
University of Connecticut	Master of Science in Business Analytics and Project Management	M
University of Dallas	MS or MBA—Business Analytics	M
University of Denver	Business Information and Analytics Certificate	C
University of Denver	Master of Science in Business Analytics	M
University of Evansville	Statistics and Data Science	B
University of Florida	Master of Science in Information Systems and Operations Management (Business Intelligence and Analytics specialization)	M
University of Georgia	MBA with concentration in Business Analytics	M
University of Illinois at Chicago	Health Informatics (IBHE-Approved Certificate)	C

School	Program	Degree
University of Illinois at Chicago	Master of Science in Health Informatics Research Track	M
University of Illinois at Springfield	Graduate Certificate in Business Intelligence	C
University of Illinois at Urbana-Champaign	Master of Computer Science in Data Science	M
University of Illinois at Urbana-Champaign	Master of Science in Statistics: Analytics Concentration	M
University of Kansas	Business Analytics	C
University of Louisville	Certificate in Data Mining	C
University of Maryland-Baltimore County	Master of Science in Information Systems	M
University of Maryland-College Park	Master of Information Management	M
University of Maryland-College Park	MS in Business, Marketing Analytics	M
University of Maryland-College Park	Online MBA—Specialization in Information Systems and Business Analytics	M
University of Maryland-University College	Master of Science in Data Analytics	M
University of Massachusetts Amherst	Certificate in Data Science	C
University of Massachusetts Amherst	MS with Data Science Concentration	M
University of Memphis	Master's Program in Bioinformatics	M
University of Miami	Master of Science in Business Analytics	M
University of Michigan-Ann Arbor	Data Science	B
University of Michigan-Ann Arbor	Graduate Data Science (DS) Certificate Program	C
University of Michigan-Dearborn	Master of Science in Business Analytics	M
University of Minnesota	Business Analytics	M
University of Minnesota-Duluth	Retail Marketing Analytics	B
University of Missouri-St Louis	Graduate Certificate in Business Intelligence	C
University of Montana	Certificate in Big Data Analytics	C
University of Nebraska at Omaha	Data Science Concentration	B
University of Nebraska at Omaha	Data Science Concentration	M
University of Nevada-Reno	Information Systems Data Analytics Track	M
University of New Haven	M.B.A.—Business Intelligence Concentration	M
University of North Carolina at Chapel Hill	Graduate Certificate in Public Health Informatics	C

School	Program	Degree
University of North Carolina at Charlotte	Computing and Information Systems	D
University of North Carolina at Charlotte	Graduate Certificate in Data Science and Business Analytics	C
University of North Carolina at Charlotte	Graduate Certificate in Health Informatics	C
University of North Carolina at Charlotte	M.S. in Mathematics with Concentration in Applied Statistics	M
University of North Carolina at Charlotte	Professional Science Master's (PSM) in Data Science and Business Analytics (DSBA)	M
University of North Carolina at Greensboro	Graduate Certificate in Business Analytics	C
University of North Texas	MS—Business Analytics	M
University of Notre Dame	Master of Science in Applied and Computational Mathematics and Statistics with a specialty in Applied Statistics	M
University of Notre Dame	Master of Science in Business Analytics	M
University of Notre Dame	Master of Science in Data Science	M
University of Oklahoma Norman Campus	Master of Science in Engineering—Data Science and Analytics Emphasis	M
University of Oklahoma Norman Campus	MS in Management Information Systems	M
University of Pittsburgh-Bradford	Master's Degree in Biomedical Informatics	M
University of Pittsburgh-Bradford	Master's Degree in Biostatistics	M
University of Pittsburgh-Pittsburgh Campus	MSIS—Big Data Analytics	M
University of Rochester	Data Science	B
University of Rochester	MA and MS in Medical Statistics	M
University of Rochester	Master of Science in Data Science	M
University of Rochester	MS in Business Administration (Concentration in Business Analytics)	M
University of Rochester	Professional MS in Computer Science (Concentration in Data Analytics)	M
University of Saint Joseph	Graduate Certificate in Health Informatics	C
University of San Francisco	Data Science	B
University of San Francisco	Master of Science in Analytics	M
University of San Francisco	Master of Science in Health Informatics	M
University of South Carolina-Columbia	Master of Applied Statistics	M
University of South Florida Sarasota-Manatee	Business Analytics Certificate	C

School	Program	Degree
University of South Florida-Main Campus	Analytics and Business Intelligence Certificate	C
University of South Florida-Main Campus	MS in Health Informatics Online	M
University of Southern California	Master of Science in Business Analytics	M
University of Southern California	Master of Science in Computer Science—Data Science	M
University of Southern California	Master of Science in Healthcare Decision Analysis Concentration in Business Intelligence	M
University of St Francis	Graduate Certificate in Business Analytics	C
University of St Francis	MBA with a concentration in Business Analytics	M
University of St. Thomas	M.S. in Data Science	M
University of the Pacific	Master of Science in Analytics	M
University of Utah	Big Data Certificate	C
University of Utah	MS in Computing: Data Management and Analysis	M
University of Virginia	Master of Science in Data Science (MSDS)	M
University of Washington-Seattle Campus	Certificate in Data Science	C
University of Washington-Seattle Campus	M.S. in Data Science	M
University of Washington-Tacoma Campus	Big Data	D
University of Washington-Tacoma Campus	Certificate in Business Analysis	C
University of Washington-Tacoma Campus	Certificate in Business Intelligence	C
University of Wisconsin Colleges	Graduate Certificate in Business Analytics (UW-Milwaukee)	C
University of Wisconsin Colleges	Online Master of Science in Data Science	M
University of Wisconsin-Madison	Master of Engineering in Applied Computing and Engineering Data Analytics	M
University of Wisconsin-Madison	Master of Science in Statistics—Data Science	M
University of Wisconsin-Milwaukee	Online Graduate Certificate in Business Analytics	C
University of Wisconsin-River Falls	Data Science and Predictive Analytics	B
Valparaiso University	Bachelors of Arts in Business Analytics	B
Valparaiso University	Bachelors of Science in Data Science	B
Valparaiso University	Masters of Science in Analytics and Modeling	M
Villanova University	Graduate Certificate in Applied Statistics	C
Villanova University	Master of Science in Analytics	M
Villanova University	Master of Science in Applied Statistics	M
Villanova University	Online Master of Business Administration	M

School	Program	Degree
Virginia Commonwealth University	Master of Science in Business, concentration in Decision Sciences and Business Analytics	M
Virginia Polytechnic Institute and State University	Computational Modeling and Data Analytics	B
Wake forest University	Masters of Science in Business Analytics	M
Wake Tech Community College	Associate in Applied Science degree in Business Analytics	A
Washington State University	Online Executive MBA	C
Washtenaw Community College	Applied Data Science Certificate	C
West Virginia University	Master of Professional Studies in Statistical and Data Sciences	M
Westminster College	Minor in Data Science	B
William and Mary	Business Analytics	M
Winona State University	Data Science	B
Worcester Polytechnic Institute	Data Science	B
Worcester Polytechnic Institute	Data Science	D
Worcester Polytechnic Institute	Graduate Certificate in Data Science	C
Worcester Polytechnic Institute	Master of Science in Data Science	M
Yale University	Statistics and Data Science	B
Yale University	Statistics and Data Science	D

Degree programs were mined from self-reports to the author, online material, university and college websites, and published reports

The manufacturer's authorised representative in the EU is Springer Nature Customer Service Centre GmbH, Europaplatz 3, 69115 Heidelberg, Germany. If you have any concerns regarding our products, please contact ProductSafety@springernature.com

Printed and bound by CPI Group (UK) Ltd, Croydon, CR0 4YY
22/09/2026
02220945-0016